Powder Metallurgy
For
High-Performance Applications

SAGAMORE ARMY MATERIALS RESEARCH
CONFERENCE PROCEEDINGS
Published by Syracuse University Press

Fundamentals of Deformation Processing
Walter A. Backofen et al., eds.
(9th Proceeding)

Fatigue–An Interdisciplinary Approach
John J. Burke, Norman L. Reed, and Volker Weiss, eds.
(10th Proceeding)

Strengthening Mechanisms–Metals and Ceramics
John J. Burke, Norman L. Reed, and Volker Weiss, eds.
(12th Proceeding)

Surfaces and Interfaces I
Chemical and Physical Characteristics
John J. Burke, Norman L. Reed, and Volker Weiss, eds.
(13th Proceeding)

Surfaces and Interfaces II
Physical and Mechanical Properties
John J. Burke, Norman L. Reed, and Volker Weiss, eds.
(14th Proceeding)

Ultrafine-Grain Ceramics
John J. Burke, Norman L. Reed, and Volker Weiss, eds.
(15th Proceeding)

Ultrafine-Grain Metals
John J. Burke and Volker Weiss, eds.
(16th Proceeding)

Shock Waves and the Mechanical Properties of Solids
John J. Burke and Volker Weiss, eds.
(17th Proceeding)

Powder Metallurgy For High-Performance Applications

EDITORS

JOHN J. BURKE

Planning Director, Army Materials and Mechanics Research Center

VOLKER WEISS

Professor, Syracuse University

Proceedings of the 18th Sagamore Army Materials Research Conference. Held at Sagamore Conference Center, Raquette Lake, New York, August 31-September 1-3, 1971. Sponsored by Army Materials and Mechanics Research Center, Watertown, Mass., in cooperation with Syracuse University. Organized and directed by Army Materials and Mechanics Research Center in cooperation with Syracuse University.

SYRACUSE UNIVERSITY PRESS 1972

Library of Congress Cataloging in Publication Data

Sagamore Army Materials Research Conference, 18th,
Raquette Lake, N. Y., 1971.
Powder metallurgy for high-performance applications.

(Sagamore Army Materials Research Conference.
Proceedings, 18)
Sponsored by Army Materials and Mechanics Research
Center, Watertown, Mass., in cooperation with Syracuse
University.
Includes bibliographical references.
1. Powder metallurgy—Congresses. 2. Metal
powder products—Congresses. I. Burke, John J., ed.
II. Weiss, Volker, 1930-   ed. III. United States.
Army Materials and Mechanics Research Center.
IV. Syracuse University. V. Title. VI. Series.
UF526.3.S3 no. 18 [TN695]   623'.028s [671.3'7]
ISBN 0-8156-5036-1   72-5215

Printed in United States of America
Composed and printed by Science Press, Inc., Ephrata, Pa.
Bound by Vail-Ballou Press, Inc., Binghamton, N.Y.

Sagamore Conference Committee

Chairman
John J. Burke, Army Materials and Mechanics
Research Center

Secretary
Sunil K. Dutta, Army Materials and Mechanics
Research Center

Program Director
Volker Weiss, Syracuse University

Program Committee
Dr. John J. Burke, Army Materials and Mechanics
Research Center
Prof. Nicholas J. Grant, Massachusetts Institute
of Technology
Prof. Fritz V. Lenel, Rensselaer Polytechnic
Institute
Mr. Harold Markus, Frankford Arsenal
Mr. Frank Zaleski, Frankford Arsenal
Mr. Lewis R. Aronin, Army Materials and Mechanics
Research Center
Prof. Alan Lawley, Drexel University
Dr. James K. Magor, Army Research Office, Durham

Arrangements at
SAGAMORE CONFERENCE CENTER
John Lathrop, Syracuse University

Foreword

The Army Materials and Mechanics Research Center has conducted the Sagamore Army Materials Research Conferences in cooperation with the Metallurgical Research Laboratories of the Department of Chemical Engineering and Metallurgy of Syracuse University since 1954. The purpose of the conferences has been to gather together scientists and engineers from academic institutions, industry, and government who are uniquely qualified to explore in depth a subject of importance to the Army, the Department of Defense, and the scientific community.

The recognition of the impact of the powder-metallurgy process for high-performance applications has resulted in significant advances in the technology of powder production, compaction, sintering, and deformation processing.

This volume, "Powder Metallurgy for High-Performance Applications," addresses the broad areas of: emerging technologies in powder metallurgy; advances in the technology of powder preparation; recent developments in pressing and sintering; processing of wrought products; and high-performance applications.

The assistance provided by Dr. S. K. Dutta and Mr. E. J. Lemay of the Army Materials and Mechanics Research Center during the editing of this book is acknowledged.

The continued active interest and support of these conferences by Dr. A. E. Gorum, Director, Major K. I. Kawano, Commanding Officer, and Mr. E. N. Hegge, Associate Director, of the Army Materials and Mechanics Research Center, is appreciated.

Sagamore Conference Center   The Editors
Raquette Lake, New York
August 1971

Contents

SESSION V
PROCESSING OF WROUGHT PRODUCTS (B)
Frank Zaleski, *Moderator*

SESSION VI
HIGH PERFORMANCE APPLICATIONS
Norman B. Schwartz, *Moderator*

INTRODUCTION

MODERATOR: ALVIN E. GORUM
Director, Army Materials and Mechanics Research Center
Watertown, Massachusetts

1. Emerging Technologies in Powder Metallurgy

A. LAWLEY
Drexel University
Philadelphia, Pennsylvania

ABSTRACT

The potential of the powder-metallurgy process in high-performance applications has brought about significant advances in the technology of powder production, compaction, sintering, and deformation processing. This, in turn, has been coupled with a clearer understanding of the basic phenomena involved. Successful production of fully dense structural shapes with the desired combination of strength, ductility, and impact and fatigue resistance, requires an intimate interrelationship between all steps in the fabrication route, and input from diverse disciplines. The primary aim in the overview is to provide a background for the chapters to follow through a consideration of possible unifying aspects of the technologies involved. Particular attention is given to hot isostatic pressing and powder preform forging. The importance of structural characterization is also discussed.

Introduction

The primary goal of this first chapter is to provide a background for those which follow, to introduce the general philosophy behind the content of the book, and to delineate important objectives. A brief discussion of the timeliness of powder metallurgy, and in particular powder metallurgy for high-performance applications, as the topic of this volume, is appropriate. It is also desirable to summarize and clarify a number of working definitions, nomenclature, and semantic problems that now pervade the powder-metallurgy literature.

In the context of powder metallurgy, high performance means different things to different people. High performance may be considered in terms of any grouping of properties (e.g., physical, electrical, chemical, mechanical). The main thrust of this volume is with mechanical behavior in that high performance is equated to use of the P/M part or component as a structural member, or the state of stress is of primary importance in design criteria. Although mechanical properties are generally considered in relation to the interplay between strength

3

and ductility, particular interest in high-performance P/M materials is centered on fatigue behavior, impact strength, and creep resistance.

The Conventional Approach

The conventional powder-metallurgy route (Figure 1) involving some form of compaction and subsequent heat treatment (sintering) provides a means for producing parts exhibiting a wide range of tensile strengths comparable to wrought material. However, in almost all cases, the P/M parts are characterized by low ductility and impact strength, and questionable fatigue resistance. Structural applications are therefore extremely limited. The primary reason for these deficiencies lies in the residual porosity; typical void contents are in the range of

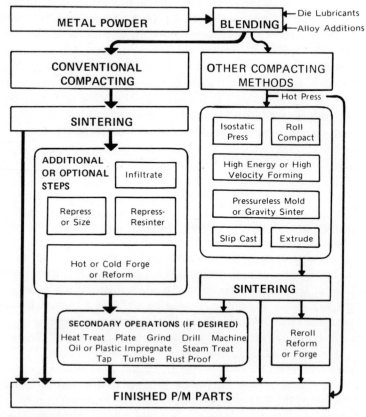

Figure 1. Basic steps in the conventional powder-metallurgy process. (Courtesy Metal Powder Industries Federation.)

5-25 percent. Improvements in property levels (fatigue, impact, ductility) have been achieved through approaches such as repressing, infiltration, post-heattreatments, or modification of the pore morphology. However, notwithstanding the cost of these extra steps, the extent of improvement in properties rarely allows for direct competition with, or replacement of, a conventional cast and wrought material. On this basis, and allowing for a continuing improvement in powder materials (*e.g.*, purity, shape, size), compacting technology and equipment, growth of the conventional P/M market is a direct function of the extent of innovation possible in new products and markets in which high performance is not a requirement. A growth rate ~15 percent per year is predicted for the production of P/M parts; within this category, the greatest potential is for larger parts of higher density.

The wide range of property levels now attainable in semidense P/M parts, and typical applications, are summarized in Table I. A comparison with property levels five to ten years ago would reveal significant improvement; the high-density steels and nickel steels are now in everyday use. In the present context, these P/M parts (characterized by tensile strength levels $\leq 100,000$ psi and with elongations to fracture ≤ 5 percent) do not qualify as high-performance materials.

The Approach for High Performance

Clearly, a direct approach to the utilization of powder-metallurgy materials in high-performance applications lies in the achievement of full density—*i.e.*, it is necessary to eliminate all forms of porosity. Primary stimuli behind the current intense level of interest and research and development in fully dense P/M materials include: an overcapacity on the part of the powder producers (capacity ≥ 3 times the demand for iron powders) coupled with a limited market for semidense materials; the inability of conventional processing procedures to fabricate many complex alloys; production of structural components on a competitive basis vis-à-vis wrought products; and significant advances in the technology of powder production.

As might be expected, many varied approaches are being studied in the quest for high-performance levels in P/M materials. The majority of these research and development programs involve either (1) the direct and complete densification of the powder (*i.e.*, rely on improved compaction technology), or (2) follow the route of preparing a powder preform which is subsequently worked (*i.e.*, deformation-processing) to full density.

Isostatic Compaction

Isostatic compaction [1-5], and in particular hot isostatic pressing [2-4], is perhaps the most promising method for the direct densification of loose powders

TABLE I

Properties and Applications of Powder-Metallurgy Parts*

FERROUS MATERIALS

Material	Condition	Density G per CuCm	% Theo.	Tensile Strength (psi)	Elonga-tion (%)	Typical Applications
Unalloyed iron, 99.9% Fe	As-sintered	6.0 6.5 7.5	78 83 96	20,000 30,000 40,000	3.0 10 20	Structural (lightly loaded gears), magnetic (motor pole pieces), projectile driving bands, self-lubricating bearings.
	Carbonitrided 0.040 in case 0.80 C	7.5	96	100,000 to 140,000	0.5 case 20 core	Structural, wear-resisting (small levers and cams).
Iron-carbon sintered steel	As-sintered 0.8 C	6.2	79	31,000	0.5	Structural (moderately loaded gears, levers, and cams).
	As-sintered 0.8 C	6.8	87	44,000	1.0	
	Heat-treated (large parts)	6.8	87	64,000	0.5	Structural (moderately loaded gears, levers, and cams requiring wear-resistance).
Iron-copper-carbon alloy	As-sintered 5.0 Cu, 0.8 C	6.2	80	60,000	0.8	Structural (medium loads including gears, cams, support brackets, levers, and ratchets)—can be heat-treated to a high degree of wear-resistance.
	Infiltrated 20.0 Cu, 0.8 C	7.4	95	80,000	0.7	
	Infiltrated, heat-treated 20.0 Cu, 0.8 C	7.4	95	100,000	0.5	
Nickel-alloy steel	Heat-treated 4.0 Ni, 1.0 Cu, 0.70 C	6.8	87	100,000	0.5	Structural, wear-resisting, high-stress (planetary differential and transmission gears up to 6 hp).

Material	Condition	Density G per CuCm	Density % Theo.	Tensile Strength (psi)	Elongation (%)	Typical Applications
	Heat-treated 7.0 Ni, 2.0 Cu, 0.5 C	7.4	95	175,000	2.5	Structural, high-stress, impact-resistance (shifter lugs and clutches).
	Carbonitrided 0.040 in case 4.0 Ni, 1.0 Cu, 0.0 C (core)	6.9	89	90,000	3.5 (core)	Structural, wear-resisting, high-stress, and requiring welded assembly (welded assembly of pinion and sprocket).

NONFERROUS MATERIALS

Material	Condition	Density G per CuCm	Density % Theo.	Tensile Strength (psi)	Elongation (%)	Typical Applications
Copper-nickel-zinc alloy	Single-pressed as-sintered	7.3	82	28,000	7	Structural, nonacid corrosion-resisting (gears, levers, chuck jaws, parts for marine exposure).
		7.9	89	35,000	15	
Sintered brass	Prealloyed, as-sintered	7.6	87.5	29,600	10	Mechanical components, atmospheric corrosion-resisting (builders' hardware, mechanism housings, lock parts, pump housings).
	Coined and resintered	8.0	92	31,000	24	
Aluminum alloy (Alcoa Type 201 AB)	As-sintered	2.59	—	28,300	2.5	Lightly loaded gears and ratchets, camera parts, circuit board, heat sink, cabinet hardware.
	Sintered and coined	2.63	—	32,200	3.0	
	Sintered, coined, heat-treated	2.63	—	49,400	2.0	

*Adapted from a compilation by S. H. McGee in *Metals Progress* (April 1971).

or cold compacts. By the simultaneous application of temperature and pressure it is feasible to achieve 100 percent density or close to full density in a wide range of hard-to-work materials. High-speed tool steels, superalloys, beryllium, and the refractory metals are particularly amenable to hot isostatic pressing. To clarify terminology, the following descriptive titles all refer to hot isostatic pressing: hot isostatic compaction, gas-pressure bonding, gas-pressure compaction, gas-pressure consolidation, and pressure bonding.

While extensive mechanical property correlations are lacking, it is usual to find that these properties are superior to those of the same alloy after casting. In some instances, e.g., nickel-base superalloys, hot isostatically compacted material may be subsequently worked either hot or cold without failure. This is in comparison to the nonworkable nature of the cast alloy. Key to this improvement in ductility lies in the uniformity of structure, isotropic behavior, and small grain size following hot isostatic pressing.

A recent innovation in compaction, in which a uniaxial stress and an isostatic state of stress are combined, appears to offer further potential for densification [6]. Developed and proven for cold compaction, the approach can be utilized at elevated temperatures. Briefly, in this "combined stress state" compaction, shear stresses are developed within the compact of magnitude:

$$\tau = \frac{(\sigma_1 - \sigma_3)}{2} \tag{1}$$

where σ_1 is the axial stress and σ_3 is the isostatic stress. The enhancement in densification is illustrated in Figures 2 and 3 for sponge iron and aluminum, respectively. In effect, the procedure increases the performance of the press; at 12,000 psi isostatic pressure plus a shear stress of 30,000 psi, sponge iron is compacted to 80-percent density. Using only isostatic pressure, 40,000 psi is necessary to achieve the same density. Of primary interest is the limiting density that can be achieved. The controlling mechanism is considered to be particle reorientation brought about by the imposed shear stresses. It is interesting to note that this concept of combining states of stress is commonly used in soil mechanics [7,8]. Since the stress state of the specimen is completely defined, the compaction sequence can be followed by means of the Mohr's circle representation. It is therefore possible to optimize the stress path (i.e., the sequencing and magnitudes of the increments of hydrostatic and axial stresses) in order to maximize density.

Deformation Processing of Powder Preforms

The other general approach to the achievement of full density involves the preparation of an intermediate porous powder preform. Densification of the

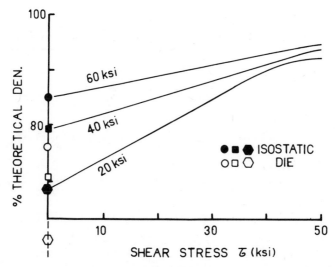

Figure 2. A comparison of die compaction, isostatic compaction, and "combined stress state" compaction on the densification of sponge-iron powder (MH-100).

preform is then brought about by a hot or cold working operation. The procedure is illustrated schematically in Figure 4. The porosity required in the preform is a function of the subsequent mode of deformation and may be as high as 30 percent or <1 percent. Most preforms have been prepared by die-compac-

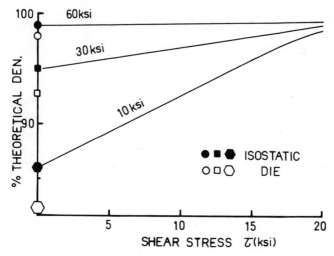

Figure 3. A comparison of die compaction, isostatic compaction, and "combined stress state" compaction on the densification of aluminum powder (201 AB).

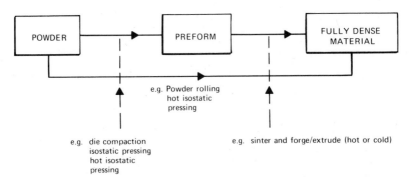

Figure 4. Basic steps in the production of, a fully dense material via an intermediate powder preform.

tion, cold isostatic pressing, hot isostatic pressing, or a combination of compaction steps. The most frequently used working operations are forging, extrusion, or rolling at ambient or elevated temperatures. A typical sequence for tool steels [9,10] involves hot isostatic pressing followed by hot rolling or forging to 100-percent density. For nickel-base superalloys, hot extrusion [11-16] or hot forging [17,18] produces the desired densification, while in most iron-base alloys forging is most common [19]. In aluminum-base alloys, emphasis is on the forging [20] or extrusion [21] of preforms. The final densification step in titanium alloys involves hot extrusion or forging [22,23].

Of these emerging technologies based on preform processing, that of preform forging [19,24] appears set to make a major impact in high-performance P/M materials. It is therefore appropriate to give some consideration to property levels and applications. Important applications (actual, prototype, and projected) in the aerospace and automobile industry are listed in Table II. Property

TABLE II

Potential Areas of Application—Preform Forgings

AEROSPACE

Component	Material
Turbine blades; trunnion rings; starter rotor blades; turbine discs; inlet guide vanes.	410 stainless; modified 440 stainless; Ti-6Al-4V; Inconel 718; U-700 and other superalloys.

AUTOMOTIVE

Component	Material
Manual transmission gears; drive-shaft flanges; connecting rods; side gear and pinion; alternator poles; rings gears.	1010, 1040, 1050 steels; 4000, 4400 low-carbon steels; 4630; 8630.

TABLE III

Representative Mechanical Properties of Preform Forged Powder Materials

Material	Yield Strength (ksi)	Tensile Strength (ksi)	Elongation (%)	R.A. (%)	Impact at 20°C (ft. lb.)
1040 Steel	72	>80<	>25<	>40<	25-30
Fe-Ni-Mo	–	80-150	12-20	45-50	20-30
Fe-Mn-Ni-Cr	–	60-130	9-30	15-60	6-65
70/30 Brass	33	55	30	–	–
Aluminum	6.3	14	39	–	–
Ti-6A1-4V	165	–	8	–	–
Inconel 713C	186	–	10	–	–

ranges available for a number of materials are summarized in Table III. These property levels represent a tremendous improvement and advance over those characteristic of conventional compacted and sintered powders. For iron-base powders, tensile strengths ~300,000 psi with associated elongations to fracture ~12 percent are now commonplace at full density. Where available, impact and fatigue properties are found to be comparable to those of the wrought material [25].

A clarification of terminology is needed in relation to the forging of powders or powder preforms. The terms repressing, sizing, coining, and restriking, whether performed hot or cold, are used in the context of preform forging. However, in these operations, the major event is that of densification; only a limited amount of material flow occurs (e.g., localized flow in coining). The connotation of these terms with conventional forging is therefore misleading since forging involves significant amounts of material flow. Confined- and closed-die preform forging simulate conventional forging. Other suggested descriptions of the process include hot recompacting and sinter-forging. The latter term has validity if the preform is cooled after sintering and subsequently reheated for the forging step.

Design of the preform forging process to take full advantage of the beneficial features and overcome detrimental aspects requires careful consideration of the interaction of the preform material characteristics and the deformation during forging. Specifically, these are the powder characteristics (purity, size, shape, inclusions), preform (porosity, uniformity, structure, shape), and forging operation (die design, lubrication, coating, temperature). Obviously, similar considerations apply to extrusion or other forms of preform processing. In some cases, it is advantageous to consider the preform material as a two-phase material consisting of the matrix metal or alloy and the pores. In other instances, it is convenient and necessary to think in terms of a quasicontinuum material. It is clear that optimization of the process requires a complete understanding of the mechanics of deformation of a body of changing density, and the structural characteristics associated with all stages of the process.

Application of ductile fracture criteria developed for conventional materials, and the use of a quantitative analysis of the mechanics of deformation of semi-dense materials, provide a semiquantitative picture of flow and fracture in preform forging. In turn, this establishes guidelines for design of the total process sequence.

Ideally, the ultimate goal of studies of preform forging should be the establishment of a fail-safe criterion [26]. This is best illustrated by reference to Figure 5, which relates the preform design (height H_o : diameter D_o) to the shape

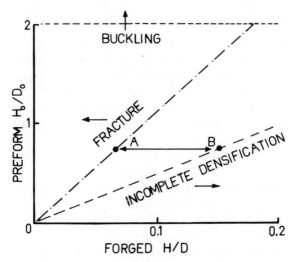

Figure 5. The "forging corridor" for the design of the preform in the powder preform forging process.

of the final forging (height H : diameter D). For a given preform geometry (*e.g.*, $H_o/D_o = 0.75$) the diagram provides the range of H/D (*i.e.*, between points A and B) for the final forging which will give rise to densification but will not fracture. Thus, the area between the limiting lines for "fracture" and "incomplete densification" represents a "forging corridor". Clearly, the relative location of these two lines will be a function of the material, and the forging conditions; a specific relationship between preform geometry and geometry of the final forging is required for each particular situation. It is entirely possible that the line of incomplete densification lies to the left of the fracture line. Under this condition, the diagram gives the geometry (H/D) of the forging at the onset of fracture; at this point the deformation must be stopped, and the necessary further densification achieved under conditions of repressing with minimal amounts of lateral flow of material. The extension of this form of approach to the design of other powder-preform processing sequences is a prerequisite to their full utilization in the production of high-performance materials.

Structure-Property Considerations

With increasing emphasis on high or full density and with properties comparable to those of the wrought state, it becomes advantageous and important to conduct detailed structural studies on P/M materials. Historically the characterization of structure in compacted and sintered materials involved optical metallography, and the replication of powder particles. Until recently, the application of transmission electron microscopy was limited to *In-situ* studies of sintering [27-29]. It is now possible to characterize substructure in a manner analogous to that of wrought materials. Scanning electron microscopy has provided a further stimulus to the detailed examination of surface features in powders [30-33] or compacts [31,32]. Because of the importance of the structure-property relationship in high-density materials, a brief consideration is given to some recent observations; these show clearly the potential and value of the various techniques.

Scanning Electron Microscopy

The ability of the scanning electron microscope to characterize powder-particle morphology (*e.g.*, size, shape, surface structure, pore detail) is illustrated in Figure 6. These powders reflect the mode of formation. MH-100 is an irregular porous iron powder resulting from the reduction of the oxide, Figure 6(a); the nickel (F210), Figure 6(b) is an electrolytic powder with a flaky appearance. In contrast, the atomized Fe-C-Mn powder [Figure 6(c)] and the atomized prealloyed Ni-Fe [Figure 6(d)] are roughly spherical in shape.

In addition to use in powder characterization, scanning electron microscopy can be utilized in structural studies of green and sintered powder compacts. As an example, changes in the structure and geometry of particles and pores have been followed in sintered nickel and nickel-iron compacts as a function of compressive (upsetting) strain [32]. Representative sections parallel to the compression axis in the nickel compacts are given in Figure 7. The powder used was the electrolytic flake shown in Figure 6(b). It is clear that particle dimensions are reduced in the axial direction of loading and increased in radial directions; there is an attendant decrease in interparticle porosity. The flattening and elongation of grains and pores with increased axial strain is evident in Figures 7(b) and 7(c). These structural observations, together with a concurrent determination of den-

Figure 6. Scanning electron micrographs of various metal powders: (a) reduced sponge-iron powder; (b) annealed electrolytic nickel powder; (c) air-atomized iron powder (RZ365MM); (d) water-atomized nickel-iron powder (NF-1).

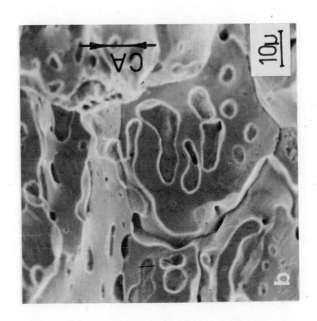

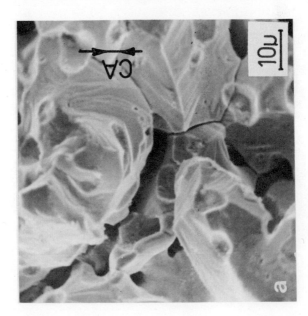

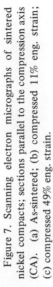

Figure 7. Scanning electron micrographs of sintered nickel compacts; sections parallel to the compression axis (CA). (a) As-sintered; (b) compressed 11% eng. strain; (c) compressed 49% eng. strain.

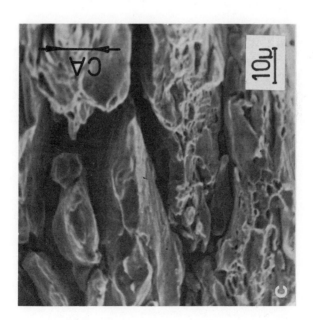

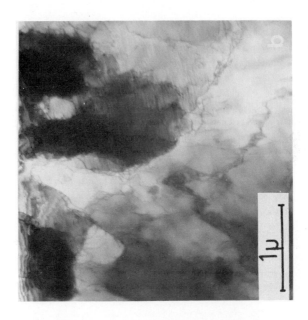

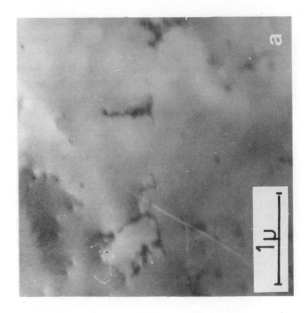

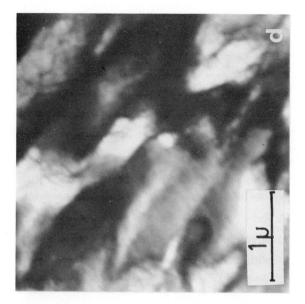

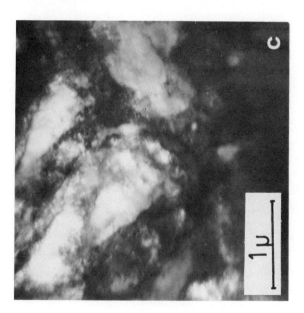

Figure 8. Transmission electron micrographs of sintered nickel compacts: (a) as-sintered; (b) compressed 10% eng. strain; (c) compressed 22% eng. strain; (d) compressed 36% eng. strain.

sity and Poisson's ratio as a function of strain, allow for a detailed understanding of the mode and mechanism of densification. Similar studies on sponge iron [31] have shown that the extent and limit of pore flattening (and therefore densification) is controlled by the orientation of cylindrical pores relative to the axis of compression.

Transmission Electron Microscopy

The preparation of electron-transparent specimens of P/M materials is usually difficult. However, techniques do now exist which allow for characterization of the dislocation substructure [31,32] in compacts of $\gtrsim 85$ percent of theoretical density. Monitoring the changes in dislocation density and dislocation substructure during densification allows for a determination of the onset of plastic flow in the powder particles, as opposed to particle reorientation and pore closure in the early stages of deformation. Examples of the development of a dislocation substructure in sintered nickel compacts are illustrated in Figure 8. The study shows that at low strains, a heterogeneous dislocation distribution exists from particle to particle. At strains $\gtrsim 10$ percent a cell structure develops, the cell size decreasing with increasing strain.

Objectives

A primary and obvious objective of this book will be to update advances and developments in each of the areas critical to the achievement of properties consistent with high-performance applications. In light of the potential of high-performance P/M materials, and recognizing the general air of optimism in the materials community, it is hoped the book will provide a springboard conducive to discussion devoid of some of the proprietary overtones that have been a characteristic of some segments of the powder-metallurgy industry.

Of the other ingredients that make up the total powder-metallurgy scene, and that fall within the scope of this book, it is hoped that serious consideration will be given to the following: (a) the delineation of key problem areas in relation to high-performance P/M materials; (b) the approach to these problem areas, *i.e.*, industry-government-university liaison and relationships vis-à-vis research and development; (c) the need for interdisciplinary studies in order to broaden and strengthen the scientific and engineering base of powder metallurgy; (d) the roles of the powder metallurgist, the parts producer, the "structure-property" oriented physical metallurgist, and the metalworking technologist in light of the peculiar and unique demands of high performance.

Finally, it should be recognized that emergence of the technologies leading to high-performance P/M materials has been keyed to improvements in powder production. Further improvements leading to specific and stringent property combinations will undoubtedly be required. It is therefore timely to examine the relative roles of empiricism and science in powder atomization.

Acknowledgments

The author acknowledges frequent and stimulating discussion with fellow faculty, in particular R. M. Koerner and H. A. Kuhn. Broad-based research in powder metallurgy at Drexel University is sponsored by DOD under project THEMIS, with technical liaison through the Metallurgy Research Laboratory, Frankford Arsenal, Philadelphia, Pennsylvania. The assistance and guidance of H. Markus and F. Zaleski is deeply appreciated.

References

1. Morgan, W.R. and Sands, R.L., "Isostatic Compaction of Metal Powders," *Metallurgical Reviews*, No. 134, *Metals Mater.* 3 (1969), 85.
2. Carmichael, D.C., "Gas Pressure Bonding Techniques," *Techniques of Metals Research*, Vol. 1, Pt. 3, R.F. Bunshah, ed., New York: John Wiley and Sons (1968), 1739.
3. Khol, R., "Isostatic Pressing," *Mach. Des.*, 42 (1970), 166.
4. Boyer, C.B., Orcutt, F.D. and Hatfield, J.E., "Hot Isostatic Bonding and Compaction Developments," *Ind. Heat.*, 37 (1970), 50.
5. Weisert, E.D. and Schwarzkopf, P.E., "Isostatic Pressing: A Three Dimensional Process," *Metal Progr.*, 99 (1971), 71.
6. Koerner, R.M. and Quirus, F.J., "High Density P/M Compacts Utilizing Shear Stresses," *Int. J. Powder Met.*, 7 (1971), 3.
7. Bishop, A.W. and Henkel, D.J., *The Measurement of Soil Properties in the Triaxial Test*, 2d ed., London: Edward Arnold Ltd. (1962).
8. Vesic, A.S. and Clough, C.W., "Behavior of Granular Materials Under High Stresses," *ASCE J. Soil Mech. Found. Div.*, 94 (1968), 661.
9. "Better Tool Steels by Powder Metallurgy," *Precis. Metal*, 29 (1971), 31.
10. Zander, K., "The ASEA-Stora Process—Production of Highly Alloyed Quality Steels by a New Quintas Process," *Powder Met. Int.*, 2 (1970), 129.
11. Friedman, G. and Kosinski, E., "High Performance Material From a Hot-Worked Superalloy Powder," *Progress in Powder Metallurgy*, Vol. 25, New York: MPIF (1969), 1.
12. Benjamin, J.S., "Dispersion-Strengthened Superalloys by Mechanical Alloying," *Metallurgical Trans.*, 1 (1970), 2943.
13. Gorecki, T.A. and Friedman, G.I., "Extended Structural Shapes from Superalloy Powders," WESTEC Conference, Los Angeles, California, March 1971.
14. Parikh, N.M., "Properties of Superalloys Made by Hot Working of Prealloyed Powders," Sixth Plansee Seminar on High-Temperature Materials, Reutte, Austria, 1968.
15. Friedman, G.I. and Lowenstein, P. "Processing Techniques for the Extrusion of Superalloy Powders," Whittaker Corporation, Nuclear Metals Division, West Concord,

Massachusetts, Air Force Materials Laboratory, Contract Report, AFML-TR-68-321 Final Report, October 1968 (AD 845-185).

16. Reichman, S.H. and Smythe, J.W., "Superplasticity in P/M IN-100 Alloy," *Int. J. Powder Met.*, 6 (1970), 65.

17. Barker, J.F. and Calhoun, C.D., "AF95 Powder Manufacturing Techniques," General Electric Company, Cincinnati, Ohio, Air Force Materials Laboratory Contract Report, AFML-TR-70-314 Final Report, December 1970 (AD 881-272L).

18. Bufferd, A.S. and Gummeson, P.U., "Application Outlook for Superalloy P/M Parts," *Metal Progr.*, 99 (1971), 68.

19. Bockstiegel, G. and Svenson, O. "The Influence of Lubrication Die Material and Tool Design Upon Die-Wear in the Compacting of Iron Powders," and Brown, G.T. and Jones, P.K., "Experimental and Practical Aspects of the Powder Forging Process," International Powder Metallurgy Conference, New York 1970, in *Modern Developments in Powder Metallurgy*, Vol. 4, H.H. Hausner, ed., New York: Plenum Press (1971), 87, 369.

20. Cebulak, W.S. and Truax, D.J., "Program to Develop High Strength Aluminum Powder Metallurgy Product: Phase I—Process Optimization," Alcoa Research Laboratories, New Kensington, Pennsylvania, Frankford Arsenal Contract Report, March 1971 (AD 882-137L).

21. Alcoa Research Laboratories, Preliminary information, Alcoa Research Laboratories Contract with Frankford Arsenal (see Ref. 20).

22. Friedman, G.I., "Titanium Powder Metallurgy," *Int. J. Powder Met.*, 6 (1970), 43.

23. Peebles, R.E., "Titanium Powder Metallurgy Forging," General Electric Company, Cincinnati, Ohio, Air Force Materials Laboratory Contract Report AFML-183-8 (V), Interim Report, February 1971 (AD 880-287).

24. Hausner, H. H., "Powder Metallurgy Forming—A Process Evaluation and Bibliography," Philadelphia: The Franklin Institute Research Laboratories, 1970.

25. Pietrocini, T.W. and Gustafson, D.A., "Fatigue and Toughness of Hot Formed Cr-Ni-Mo and Ni-Mo Prealloyed Steel Powders," *Int. J. Powder Met.*, 6 (1970), 19.

26. Downey, C.L. and Kuhn, H.A., "P/M Preform Design for Hot Forging," ASM Fall Meeting, October 1971, Detroit, Michigan.

27. Olsen K.H. and Nicholson, C.C., "Anomalous Sintering Behavior as Revealed by In Situ Electron Microscopy," *J. Am. Ceram. Soc.*, 51 (1968), 669.

28. Gessinger, C.H., Lenel, F.V. and Ansell, C.S., "Continuous Observation of Sintering of Silver Particles in Electron Microscope," *Trans. ASM*, 61 (1968), 598.

29. Kaufman, S.M., Whalen, T.J., Sefton, L.R. and Eichen, E., "The Utilization of Electron Microscopy in the Study of Powder Metallurgical Phenomena," *Advanced Experimental Techniques in Powder Metallurgy*, J.S. Hirschhorn and K.H. Roll, eds., New York: Plenum Press (1970), 1.

30. Johari, O., "Scanning Electron Microscopy in Powder Metallurgy," *Advanced Experimental Techniques in Powder Metallurgy*, J.S. Hirschhorn and K.H. Roll, eds., New York: Plenum Press (1970), 1.

31. Lawley, A., "Powder Metallurgy Production of Structural Shapes," Drexel University, Philadelphia, Pennsylvania, Project THEMIS Report for Frankford Arsenal, Technical Report 86-00-96, Progress Reports, October 1970 and May 1971 (AD 883-620L).

32. Gaigher, H.L., Koczak, M.J. and Lawley, A., "Structural Aspects of Densification in Nickel and Iron-Nickel Compacts," *Electron Microscopy and Structure of Materials*, G. Thomas, ed., Berkeley: University of California Press (1972), 389.

33. Chao, Hung-Chi, "Characterization of Commercial Metal Powders with Scanning Electron Microscope," International Powder Metallurgy Conference, New York, July 1970, in *Modern Developments in Powder Metallurgy*, Vol. 5, *Materials and Properties*, H.H. Hausner, ed., New York: Plenum Press (1971), 369.

SESSION II

MODERATOR: NICHOLAS J. GRANT
Massachusetts Institute of Technology
Cambridge, Massachusetts

2. High-Pressure Water Atomization

P. U. GUMMESON
Whittaker Corporation
West Concord, Massachusetts

ABSTRACT

Of the many commercial methods used to produce metal powders, high-pressure water atomization is economically one of the most important. It is suitable for a range of metals and alloys, and is essentially suitable for both small and large tonnages. For this and other reasons, it is perhaps the most universal of powder-manufacturing methods.

Its greatest economic and technical impact has been in the production of low- and high-alloy steels, including the stainless steels, but it is not likely to become applicable to oxidation-sensitive metals and alloys.

The process permits very good control over powder characteristics, particularly powders intended for cold compaction. The process and its influence on powder characteristics are discussed with specific reference to the influence on powder properties by the most important operating parameters.

Various modifications of jet design are reviewed.

High-Pressure Water Atomization

The use of high-pressure water for atomization of metals is advantageous because its viscosity is considerably higher than that of gases and because it has superior quenching ability. In addition, it is inexpensive. These advantages combine to make it the preferred atomizing medium for high-melting metals and alloys, for irregular particle shapes suitable for cold compacting, and for high-volume, low-cost production.

Corrosion inhibitors may be added to the water, but its oxidizing ability makes water unsuitable for very reactive metals and alloys, superalloys, and a number of tool steels. It is however, the predominant manufacturing method for iron and low-alloy steel powders and most of the stainless grades, unless spherical powder shape is desired.

The relative simplicity of water atomization makes it equally suitable for plants with one or a few thousand tons per year capacity to large-scale production of a hundred thousand tons per year or more. In the case of iron and steel powders, the advantages of water atomization are obvious in comparison to the traditional reduction processes in that the powder-making step itself—atomization —becomes a simple, relatively inexpensive addition to the traditional ways of melting and refining the steel.

This is, therefore, the material on which water atomization has had its greatest economical and technical impact in recent times. Atomized steel powder capacity has grown rapidly and presently exceeds the combined installed capacity of all other manufacturing techniques (Figure 1).

The advantages of atomized powders and, more specifically, of high-pressure water-atomized powders, are:

1. *Freedom to Alloy.* Any alloy can be made, including new alloys and systems designed especially and solely for P/M, as long as some oxidation is either acceptable or can be removed in a subsequent annealing/reduction step.

2. *All Particles Have the Same and Uniform Composition.* This eliminates

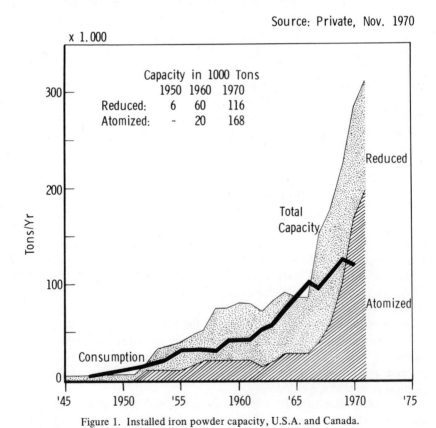

Figure 1. Installed iron powder capacity, U.S.A. and Canada.

macrosegregation and provides for more even distribution of constituents. It promotes grain-size homogeneity, and therefore workability and property reproducibility.

3. *Control of Particle Shape, Size, and Structure.* Water atomization provides reasonable control of particle shape suitable for P/M applications; however, it is not usually suitable for spheroidal and spherical shapes.

4. *Higher Purity.* Fewer nonmetallic inclusions than reduced powders; hence better compressibility, properties, and structures, especially in high- or full-density components.

5. *Lower Capital Costs.* The water-atomization process can economically be tailored to varying tonnage requirements with greater ease than most other methods.

In gas atomization of low-melting metals and alloys, true spray nozzles or integral atomizing nozzles can be used, but in the case of water atomization of high-melting alloys, the quenching effect of the water precludes any but so-called external nozzles, to prevent freeze-ups or severe errosion problems.

In the following we shall discuss the basic principles of such a system (shown in Figure 2), with special reference to the many variables that influence the end-product. Three important characteristics of metal powders are: (1) *particle size* (average particle size, size distribution, yield of useful product etc.); (2) *particle shape* (and related properties such as apparent density, flow, green strength, specific surface, etc.); (3) *particle density and structure*.

(1) *Particle Size.* For all its commercial success, atomization of high-melting metals and alloys with external nozzles is a crude process in terms of efficiency of energy utilization. The binding forces between atoms hold a mass of liquid metal together. During atomization, some of these bonds are broken in proportion to the increase in surface energy. The process must supply this energy which is proportional to the surface tension and inversely proportional to the diameter of the drops, assuming uniform particle size.

In reality, most of the applied energy in a jet system is either lost or is spent in accelerating the liquid and only a trivial amount goes into the creation of new surface energy.

The problem is therefore not one of finding enough energy to do the job, but rather of applying tensile and shear forces to the liquid on a miniscopic scale, comparable to the diameter of the particles to be produced. Another problem is that any small drops that are produced must not encounter other drops before freezing has occurred, for they will then coalesce to form bigger drops.

Finer particle sizes are favored by:

(a) low metal viscosity;
(b) low metal surface tension;
(c) superheated metal;
(d) small nozzle diameter, *i.e.*, low metal feed rate;

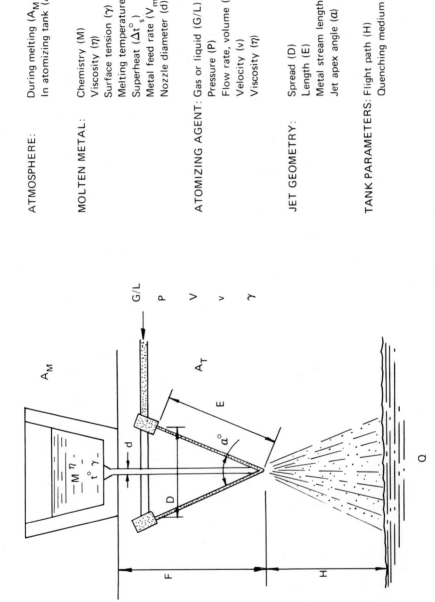

Figure 2. Major variables in the atomizing process.

(e) high atomizing pressure;
(f) high atomizing agent volume;
(g) high atomizing agent velocity;
(h) high atomizing agent viscosity;
(i) short metal stream (F);
(j) short jet length (E);
(k) optimum apex angle ($\alpha°$) (see below).

In conformity with the above tabulation, liquid atomization (*e.g.*, water) usually gives higher over-all yields of powder (*e.g.*, −100 mesh) while not necessarily producing more "fines." Commonly used pressures are 500–3,000 psi for water, and for gas, 100–400 psi. In bronze-powder atomization, K. Tamura and T. Takeda [1] achieved a comparable particle-size distribution in the −100 mesh fraction using nitrogen at 128 psi and water of 854 psi (see Table I).

TABLE I

Effect of Atomizing Pressure on Particle-Size Distribution. Metal Nozzle, 3.0 mm Diameter; Metal Temperature, 1050°C (1922°F). N_2 Atomization into Air.

Agent	Atomizing Pressure kg/cm²	psi	Particle-Size Distribution (%)					−100 Mesh Powder Yield (%)
			100/145	145/200	200/250	250/325	−335	
Water	50	711	24.7	23.8	15.4	17.9	18.2	88.3
"	55	783	22.3	22.1	15.7	19.1	20.8	91.2
"	60	854	18.6	19.3	15.9	19.5	26.8	93.7
Nitrogen	9	128	21.6	21.6	14.5	10.8	31.5	62.1

Similar work by the same authors [2] shows that with a water velocity of 340 ft/sec, changes in water volume by a factor of 6 (from 15 gal/min to 83 gal/min) did not result in any appreciably finer particle-size distribution, whereas the influence of varying water volume was somewhat more pronounced at a lower velocity (180 ft/sec).

A short metal stream is beneficial in assuring a more stable and predictable system and thereby permitting more efficient energy utilization. Short jet lengths similarly favor finer particle-size distribution since the kinetic energy decreases with the distance from the jet nozzles.

The proper jet apex angle is usually experimentally established for each system. Energy losses within a converging jet itself, at very small and very large apex angles, suggest an optimum angle for highest yield. In one study [3], over-all yield of −40 mesh superalloy powder during water atomization increased with increasing apex angles within the studied range (24−42°).

The same study [3] reports on similar relationships for two types of gas jets in superalloy atomization, a V-jet tested with apex angles of from 17-28° and an annular jet tested from 13-17°. The particle-size distribution for both jets tended to be finer as the jet angle increased. The greatest fines content (-200 mesh) resulted with the V-jet set at a 28° apex angle.

(2) *Particle Shape.* One of the advantages of atomization is the control of particle shape it permits, which is an important consideration in almost all commercial uses for metal powders. Particle shapes of atomized powders can be modified from almost perfectly spherical to highly irregular, by controlling the processes which take place in the interval between disintegration of the liquid metal stream and the solidification of the drop.

It is determined mainly by the surface tension of the molten metal and the cooling capacity, density, rate of flow, and reactivity of the atomizing agent. Gas atomization usually produces spherical particles. Water atomization usually produces irregular particles.

Sphericity (high apparent density, high flow) is favored by:

(a) high metal surface tension;
(b) narrow melting range [1];
(c) high pouring temperature (water atomizing) [4];
(d) gas atomization (slower cooling), especially inert gas;
(e) low jet velocity (water) [2,4];
(f) larger apex angles in water atomization [5];
(g) long flight paths (slower cooling) [6].

Surface tension of liquid metals is high, and a droplet once formed tends to assume spherical shape. The higher the viscosity of the atomizing medium, the greater is the deformation of the droplet. The higher the cooling rate, the shorter is the time during which the surface-tensional forces can operate to spheroidize the droplet and, therefore, the more irregular the particle shape. Impurities and alloying elements in the metal or reactions on the surface of the droplets that decrease the surface energy, will promote irregular particle shapes.

The influence of alloying additions to copper-tin and copper-tin-zinc alloys are discussed by M. Rohnisch [7]. Small quantities of phosphorous in the metal led to formation of a P_2O_5 film at the particle surfaces, which increased surface energy and resulted in the formation of spherical droplets. The existence of solid oxide films such as ZnO act in the opposite fashion, tending to give less-rounded particles. In discussing gas atomization of copper powders, Jones [8] reports that small amounts of magnesium additions cause oxide films that decrease surface energy and result in a low-apparent-density, irregular as-atomized copper powder. This has been studied by Tamura and Takeda [9], who report the most effective element to be magnesium—followed by calcium, titanium, lithium, and manganese—added in amounts up to 0.30 percent. The addition of silicon is a well-known method of influencing the particle shape of atomized stainless steel. F. Lochman [10] investigated atomization of 18-8 and 17-13

Cr-Ni steel with different Si contents. Additions of 0.5–1.0% Si to the melt resulted in spherical particles, while with 2.5–5.0% Si the particle shape was irregular.

While particle shape is not appreciably influenced by metal pouring temperature in gas atomization, it is in water-atomized powders [7]. (Inert gas atomization always produces spherical powders regardless of process variation; oxidizing gases may or may not, depending on the formation of surface oxide film.) The greater the amount of superheat, the more spherical the resultant particles.

At higher pouring temperatures, there remains enough superheat after atomization to allow surface-tension forces to create spheroids. Higher water pressures, *i.e.*, higher water velocities, result in more irregular particle shapes due to greater impact forces and to larger volumes of water with resultant more rapid quench [7].

One U.S. Patent [5] relating to a multiple-nozzle water jet, suggests that apex angles in excess of 100° tend to produce more rounded particles, while apex angles below 80° tend to produce irregular particle shapes.

By and large, however, the overwhelming influence on particle shape is related to metal surface tension and atomizing-media viscosity and quenching rate. In this vein, long flight paths between atomizing nozzle and collection liquid (if any), increases the chances of spheroidizing.

(3) *Particle Chemistry and Structure.* In the P/M of complicated alloy systems, one of the main objectives is the elimination of gross metal segregations. This concept assumes that every particle has the same analysis regardless of size. This is also generally the case for all atomizing processes, and numerous investigators have found that alloy chemistry is independent of particle size in all inert systems. (Some alloy systems may show segregations within particles, *e.g.*, Cu-Pb.) In cases where reactions can take place between the metal droplet and the atomizing agent, and/or the quenching medium, one will naturally find a greater effect on the finest particle sizes, whether the reaction be oxidation, carburization, nitriding, etc.

For obvious reasons, undesirable reactions and concurrent changes in chemistry increase with higher temperatures, slower cooling rates, and higher atomizing pressures (finer particles).

Particle structure is a function of the rate of solidification. A finer microstructure is promoted with water or liquid atomization as opposed to gas atomization, by lower metal pouring temperatures, higher atomizing agent pressure, flow rate and viscosity, and by shorter particle flight paths.

Production Atomization

Non-ferrous, ferrous, and low-alloy steels—including the stainless steels and a number of nickel alloys—can generally be air-melted and water-atomized without excessive oxidation or other detrimental effects. Such metals therefore lend themselves to high-volume, low-cost manufacturing techniques.

Conventional P/M techniques require green strength and compressibility. Atomized powders for such applications are, therefore, usually water-atomized to produce very irregular particle shapes (iron, stainless steels, etc.)

Figure 3 illustrates the principle of two simple and common jet configurations, flatstream V-jets—one a two-way plug-jet with diverging flat streams, the other a two-way curtain jet. Both of these principles are used with water as well as with gas, and are highly efficient in spite of their simplicity.

Flatstream V-jets of the type illustrated are believed to be in common use for a range of metals and alloys from iron and stainless (water) to nickel and cobalt alloys (gas). It is in fact likely that these jets account for the bulk of all atomized high-temperature metals and alloys.

For a high utilization of energy, it is obviously desirable to prevent the metal from bouncing or splashing out of the open V as it must be retained in the pocket until it has been accelerated. For this reason, "open" V-jets may require higher pressures for comparable results than the "closed" or tandem V-jets described in the following figures.

The intersection of two flatstream V-jets at 90° (as shown in Figure 4), sometimes referred to as a four-way plug-jet, creates an upside-down four-sided pyramid. This principle is further illustrated in Figure 5. In Figure 5(a), one V-jet has merely been used to close off the primary V, to help retain the metal in the pocket. In Figure 5(b), the two V-jets meet to create a well-defined apex. If the number of jets are increased, we eventually arrive at the annular jet, Figure 5(d).

In the latter, the metal stream has been shown offset from the axis of the annular jet. During water-atomizing with large amounts of water at high pressure, the pressure conditions and cooling rates within the jet may cause metal freezing inside the cone at high metal feed rates. When this happens, the jet must be shut off for an instant to release the "skull." Offsetting the metal stream from the axis of the jet helps alleviate this condition, with insignificant loss of atomizing efficiency.

While the interrupted or uninterrupted annular jet is a common design for the production of spherical powders, there would seem to be very little reason to use this design for water or other liquid jets as the much simpler plug jets are equally efficient and provide for much simpler changes of nozzles and adjustments of angles.

A high-volume water-atomization system for low-carbon, low-alloy steel powder, using a simple curtain V-jet is shown in Figure 6 [11].

This design provides for a shield around each row of metal streams to permit the introduction or a reducing gas such as hydrogen or carbon monoxide to prevent oxidation of the metal. The water level in the tank is maintained within close proximity to the nozzle assembly to further minimize oxidation.

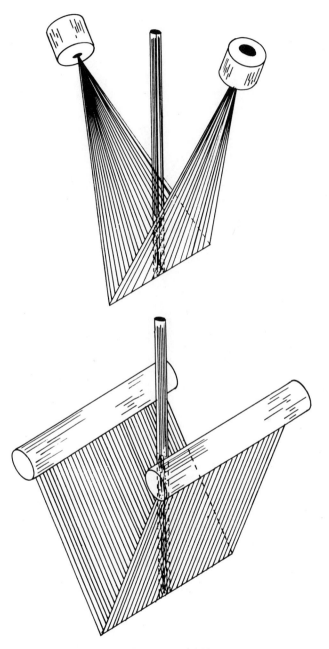

Figure 3. Flatstream V-jets.

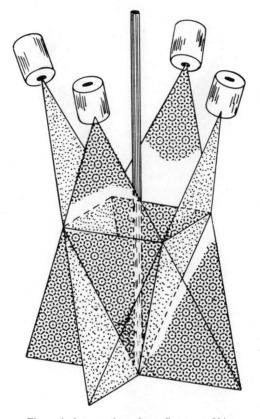

Figure 4. Intersection of two flatstream V-jets.

The metal powder is still at such a high temperature after entering the water that it tends to agglomerate, but the sloping ramps in the watertank are said to keep the powder in motion to prevent settling and mass welding.

Figure 7 shows the detail of the jet design. A later patent [12] discusses operating parameters. Metal nozzle diameters vary from 3/8" to 13/16", and the apex angle of the jet from 30° to 110°. Lower apex angles require higher water pressures, in excess of 1,000 psi. Greater apex angles permit water pressures approaching the lower limit of 500 psi.

The following conditions are said to produce as-atomized powder substantially −80 mesh with 22 percent −325 mesh:

Metal temperature	3150° F
Metal nozzle diameter	7/16"
Jet apex angle	40°
Water pressure	1,110 psi

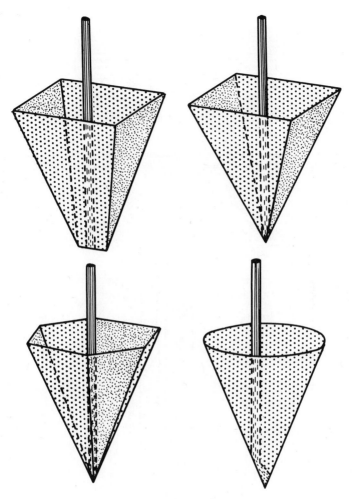

Figure 5. Jet configurations.

Water flow rate	275 gals/min
Jet nozzles	3″ long, 0.04″ opening

This powder, like most water-atomized iron and low-alloy steel powders intended for cold compacting, must be annealed.

Figures 8 and 9 show cross sections of the above-mentioned powder. It has a fine-grained microstructure. Note that the particles are dense and largely free of internal porosity and inclusions. Shape is sufficiently irregular to provide good green strength during compaction.

A competing grade of steel powder, also water-atomized and with similar

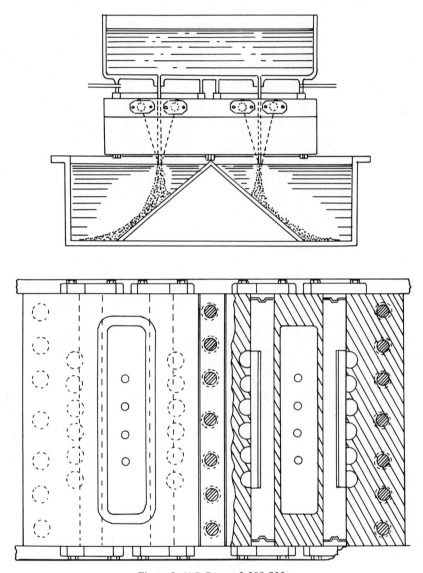

Figure 6. U.S. Patent 3,309,733.

characteristics, is shown in a scanning electron micrograph in Figure 10. The surface is highly irregular and gives the appearance of "grape clusters," a highly desirable structure from the standpoint of green strength and compactibility. It is questionable, however, how much of this agglomerated effect is really pro-

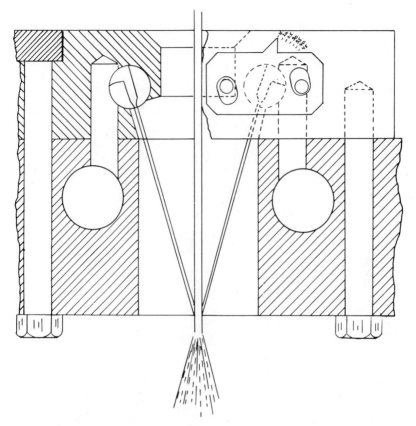

Figure 7. U.S. Patent 3,309,733.

duced in the atomization process itself and how much is a result of the subsequent annealing step.

A few words should be said about the oldest commercial processes for atomizing iron powder, *i.e.,* the atomization of a high-carbon iron melt followed by grinding and decarburizing annealing. The atomization, which is somewhat less critical as compared to the direct route from low-carbon steel, is carried out with air or water. In the former case, a partially oxidized powder is produced with a a proper balance of oxygen and carbon to permit a "self-supporting" decarburization upon heating. In the latter case, the water-atomized high-carbon powder is mixed with iron oxides to provide the necessary oxygen for subsequent decarburization. In either case, these powders show structures and characteristics somewhat between those of "pure" atomization and pure reduction of ore or millscale. Figures 11A and B and 12A, B, and C show the progressive change

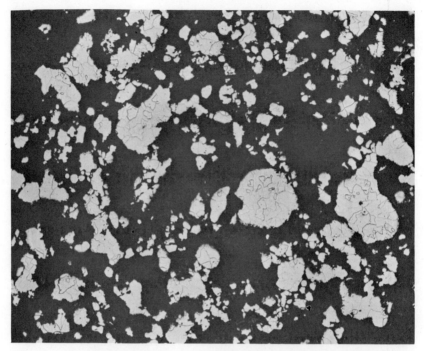

Figure 8. Atomized steel powder. 100X. (Courtesy A.O. Smith Corporation.)

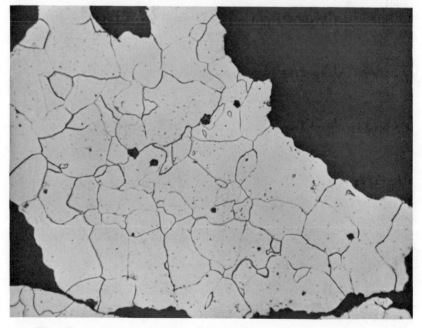

Figure 9. Atomized steel powder. 500X. (Courtesy A.O. Smith Corporation.)

Figure 10. Atomized steel powder, 0.01%C, annealed. 300X. (Courtesy Hoeganaes Corporation and Franklin Institute.)

in particle appearance from hydrogen- and carbon-reduced powders to decarburized atomized grades and a pure atomized steel powder.

Water atomization of alloys, notably stainless steels and other nickel alloys, differs little from the atomization of low-carbon steels except that these alloys sometimes are more sensitive to oxygen pickup and do not always lend themselves to subsequent economical heat treatments. Precautions must be taken to prevent excessive exposure to oxygen during atomization while the as-atomized particles must be irregular enough to provide good green strength in the as-atomized and dried condition. The atomizing chamber can be purged with any protective atmosphere prior to atomization, and the flight path should be kept short for rapid quenching. Argon or nitrogen protective atmospheres can be used. While nitrogen may be detrimental in the presence of chromium, steam can sometimes be used to advantage and is cheaper than either nitrogen or argon. Additional parameters of water atomization are illustrated by Figure 13 [5], a multiple-nozzle flatstream jet arrangement.

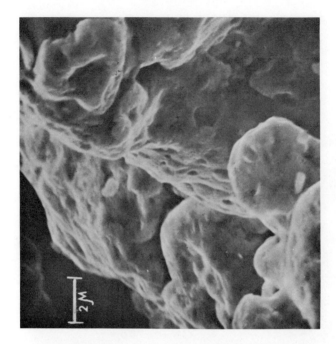

(b) Surface. X6000.

(a) Surface. X240.

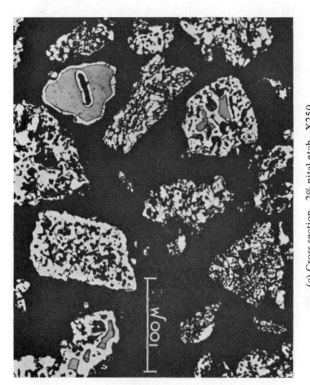

(c) Cross section. 2% nital etch. X250

Figure 11A. Hydrogen-reduced millscale. (Courtesy U.S. Steel Corporation.)

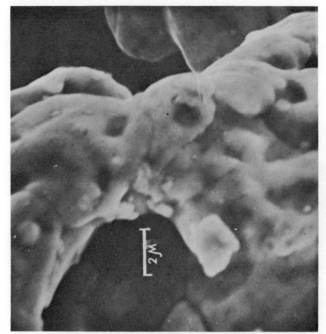

(b) Surface. X6000.

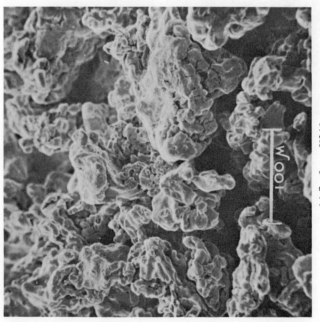

(a) Surface. X240.

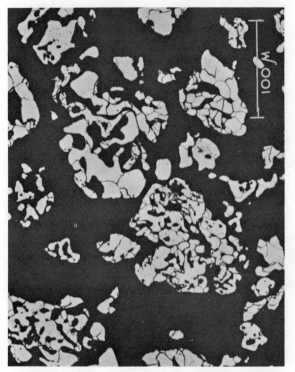

(c) Cross Section. 2% nital etch. X250

Figure 11B. Carbon-reduced ore. (Courtesy U.S. Steel Corporation.)

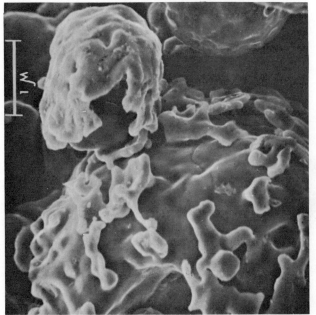

(b) Surface. X20,000.

(a) Surface. X200.

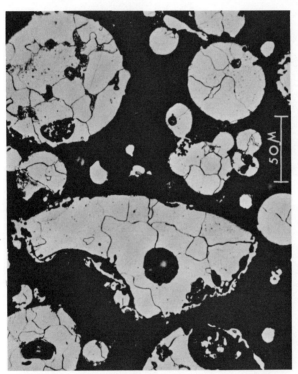

(c) Cross section. 2% nital etch. X350

Figure 12A. High-carbon iron, air-atomized and decarburized with internal oxidation. (Courtesy U.S. Steel Corporation.)

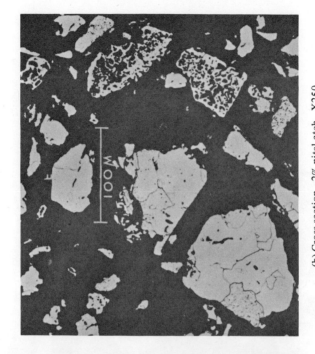

(b) Cross section. 2% nital etch. X250.

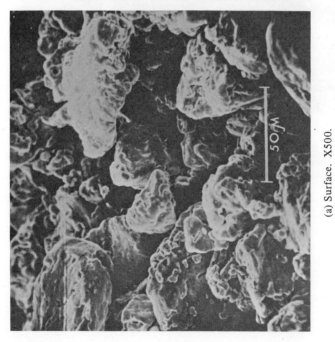

(a) Surface. X500.

Figure 12B. High-carbon iron, air-atomized and decarburized with admixed iron oxides. (Courtesy U.S. Steel Corporation.)

The jet nozzles can be rotably adjusted about their longitudinal axis; the apex angle is adjustable; and the individual nozzles can be slanted so as to produce a whirling vortex cone.

Increasing pouring rates are said to require an increasing number of nozzles, e.g., 3-6 for up to 2,700 lbs/h, up to 12-14 nozzles for over 12,000 lbs/h. Volume of water is said to range from 1/10 gal/lb metal for low-melting alloys up to 50 gal/lb for high-melting alloys.

The individual flat jet streams diverge with a total angle of about 30°, and their openings are within 4-7 inches from the axis of the metal stream to minimize energy loss.

When the flat surfaces of the liquid streams are at an angle of 90° to the metal-stream axis [Figure 13(a)], particles are claimed to approach spherical, while at 0° [Figure 13(b)] they tend to be irregular (more rapid quenching). However, with angles much less than 90°, it would seem hard to retain the material in the pocket, so that in practical operation such an adjustment is of doubtful value unless smaller angles are accompanied by increasing amounts of water.

When the apex angle is about 100° or greater, the tendency is said to be toward round particles with control of particle size decreasing, i.e., less quenching and energy utilization. At apex angles of 80° and below, the tendency is to irregular particles with more uniform particle size.

The following is quoted from the patent to illustrate the atomization of coarse, irregular 316 stainless steel with a yield of 38 percent – 100 mesh:

Metal temperature	2750° F
Metal nozzle diameter	1/4"
Metal flow rate	48 lbs/min
Jet apex angle	100° (8 nozzles)
Water pressure	1,300 psi
Water flow rate	44 gal/min
Water velocity	370 ft/sec
Angle of flat spray with metal stream	75°

While this patent is referenced to illustrate a multitude of variables, the complexity of such a system would not appear to be justified by sufficient advantages over a simple tandem V-jet as previously described.

Figure 14 shows the appearance of a molding grade 316L stainless steel powder, as water-atomized.

One method of increasing the efficiency of jets is to provide the flowing metal in thin sheets or tubular streams. One attempt in this direction [13], is shown in Figure 15, providing for a tubular metal stream intersected by one internal diverging and one external converging annular jet. With the metal components

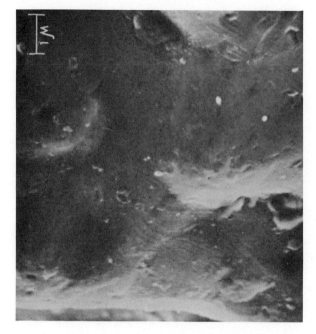

(b) Surface. X10,000

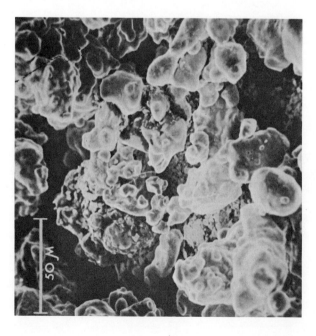

(a) Surface. X500.

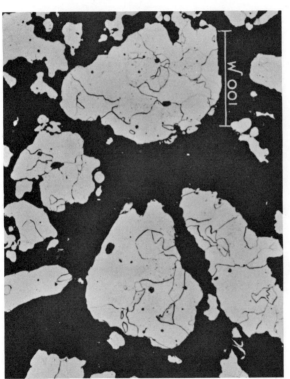

(c) Cross section. 2% nital etch. X250. (Courtesy U.S. Steel Corporation.)

Figure 12C. Low-carbon steel, water-atomized and annealed.

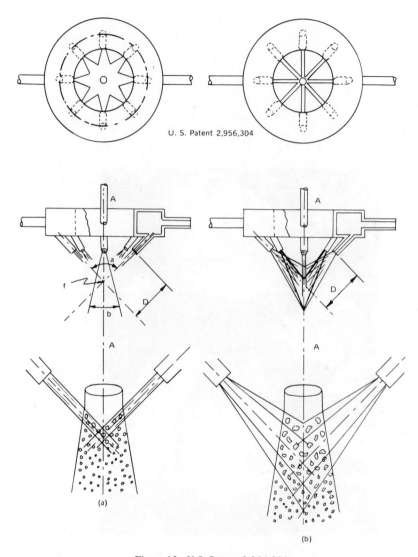

Figure 13. U.S. Patent 2,956,304.

of this design encased in refractory materials, this nozzle is said to operate successfully with high-melting alloys, with lower pressure requirement than standard nozzles.

Much undoubtedly remains to be done to advance the art of alloy atomiza-

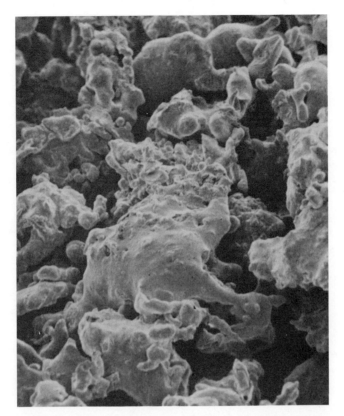

Figure 14. Water-atomized 316 stainless steel. 300X. (Courtesy Hoeganaes Corporation and Franklin Institute.)

tion, both in terms of a better theoretical understanding of the process and in terms of applying that knowledge in production. Emphasis should be placed on increasing the over-all efficiency of energy utilization. Nevertheless, atomizing techniques have already revolutionized the metal powder manufacturing industry, and are opening the door to new metallurgical materials, new P/M techniques and applications, thereby to continued growth of the P/M industry.

References

1. Tamura, K. and Tadeta, T., "On the Manufacture of Copper Alloy Powders by Liquid Atomization," *Trans. Nat. Res. Inst. Metals (Japan)*, 7 (1965).
2. Tamura, K. and Vanikawa, S., "On the Atomizing Variables in the Production of Metal Powder by Liquid Atomization," *J. Japan Soc. Powder Met.*, 15 (1968), 36.

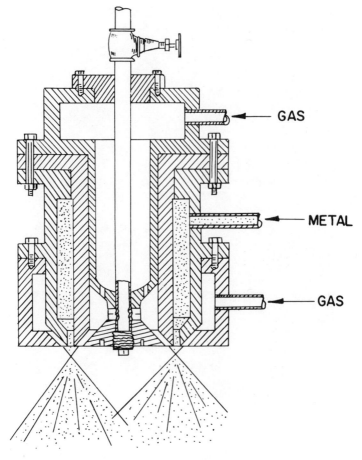

Figure 15. U.S. Patent 2,868,587.

3. Ingram, J.F. and Durdaller, C., "An Improved Manufacturing Process for the Production of Higher Purity Superalloy Powders," Hoeganaes Corporation, Riverton, New Jersey, Air Force Materials Laboratory, Contract Report AFML–TR–69–21, February 1969 (AD 853–174).
4. Small, S. and Bruce, T.J., "The Comparison of Characteristics of Water and Inert Gas Atomized Powders," *Int. J. Powder Met.,* 4 (1968), 7.
5. U.S. Patent no. 2,956,304, October 18, 1960, "Apparatus for Atomizing Molten Metal," W.L. Batten and G.A. Roberts to Vanadium Alloys Steel Company, Latrobe, Pennsylvania.
6. German Provisional Patent no. 1,027,489, 1959, G. Naser.
7. Rohnisch, M., "Production of Copper Alloy Powders by Atomization," *Neue Werkstoffe durch Pulvermetallurgische Verfahen,* Vol. 9, Ser. A., *Uber Wissenschaftliche Grundlagen der Modernen Technik,* Berlin (1964), 64.

8. Jones, W.D., *Fundamental Principles of Powder Metallurgy*, London: Edward Arnold Publishers Limited (1960), 224.
9. Tamura, K. and Takeda, T., "On the Manufacture of Irregularly Shaped Copper Powders by Gas Atomization," *J. Japan Soc. Powder Met.*, 10 (1963), 153.
10. Lochmann, F., "The Particle Shape of Atomized Powders," *Metallische Specialwerkstoffe*, Vol. 7, Ser. A, *Uber Wissenshaftliche Grundlagen der Modern Tecknik.*, Berlin: Akademic Verlag (1963), 119.
11. U.S. Patent No. 3,309,733, March 21, 1967, "Apparatus for Producing Metal Powder," B.G. Winston to A.O. Smith Corporation, Milwaukee, Wisconsin.
12. U.S. Patent No. 3,325,277, June 13, 1967, "Method of Making Metal Powder," R.A. Huseby to A.O. Smith Corporation, Milwaukee, Wisconsin.
13. U.S. Patent No. 2,868,587, January 13, 1959 "Comminuting Nozzle," W. Hegmann.

3. High-Pressure Gas Atomization of Metals

E. KLAR and W. M. SHAFER
SCM Corporation
Baltimore, Maryland

ABSTRACT

Various methods of gas atomization are described and discussed in terms of design, operating and powder characteristics, and efficiency of atomization. The description is complemented by recent laboratory data correlating powder characteristics with atomizing variables.

Introduction

This chapter presents a short review on selected aspects of gas atomization of metals. Major emphasis is placed on correlations that are of particular interest from a powder producer's point of view. This includes the technique of atomization and its efficiency, as well as certain powder characteristics such as average particle size, particle-size distribution, and particle shape. These powder characteristics critically affect such technological properties as apparent density, flow, green strength, compressibility, sinterability, and others.

Methods of Atomization

Gas and water atomization are usually classified as "two-fluid," "twin," or "pneumatic" atomization. Disintegration of the liquid to be atomized is produced by the kinetic energy of the atomizing medium.

In what is known as "external mixing" (Figure 1), contact between atomizing medium and liquid takes place outside the respective nozzles. This type of mixing is used exclusively for the atomization of metals. "Internal mixing" (Figure 1) is quite common for the atomization of materials which are liquid at room temperature.

On an industrial scale, two-fluid atomization is the most widely used method for the atomization of metals. Water atomization is generally more economical than gas atomization, and it is for this reason that gas atomization is used only

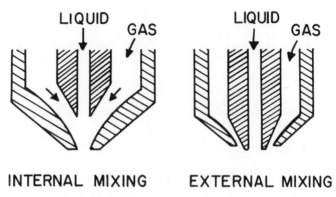

Figure 1. Two-fluid atomization designs.

where certain property specifications require its use. Thus, gas atomization—in particular, atomization with argon or nitrogen—is used where the oxygen content of a powder must be low and where spherical or approximately spherical particle shape is preferred. Rotary and vacuum atomization also compete with gas atomization for production of metal powders. Ultrasonic and vibratory atomization, to our knowledge, are at present not used on an industrial scale.

Powder Characteristics

We wish to control average particle size, particle-size distribution, particle shape, efficiency, and chemistry (homogeneity, gas content, surface quality) for technical as well as for economical reasons. Any one or several of these properties can become of primary importance depending on the application. Also, all of these properties are affected by atomization design variables as well as by atomization operating variables. Average particle size and particle-size distribution are of particular interest to the powder producer. The powders he sells frequently are subject to narrow specifications. It is not difficult to control average particle size such as the mass median diameter to within a few microns, but it is quite difficult to affect the particle-size distribution. In many cases, the particle-size distribution is an important property from both technical and economical points of view. This relationship is illustrated in Figures 2A and 2B. Assuming that the screen specification calls for a powder falling between 140 and 270 mesh, and that the as-atomized powders follow logarithmic normal distributions, which often is true, we can calculate the yield of -140 to $+270$ mesh powder as a function of the geometric standard deviation which is a measure of the dispersion of the powder. The optimum geometric mean mass diameter, d_{gm}, turns out to be 74 microns and, for the example shown in Figure 2A, the powder yield is 28 percent for $\sigma_g = 2.5$, and 36 percent for $\sigma_g = 2.0$. This constitutes a relative improvement of 29 percent. Most values for σ_g of metal pow-

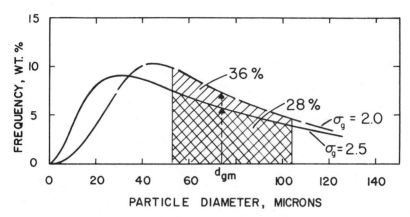

Figure 2A. Effect of dispersion on yield.

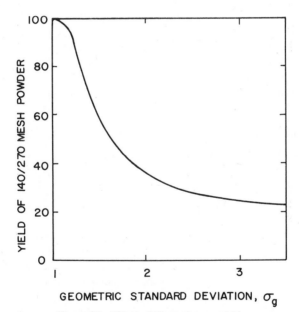

GEOMETRIC STANDARD DEVIATION, σ_g

Figure 2B. Effect of dispersion on yield.

ders produced by twin-fluid atomization and as reported in the literature fall within the range of 3 to 2. Thus, as shown on Figure 2B, the improvement in yield accelerates with decreasing σ_g.

In a recent study on the atomization of bronze, Nichiporenko [8] found that increasing gas velocity or decreasing diameter of the metal stream produced narrower particle-size distributions. On theoretical grounds, we can postulate the following nozzle-design characteristics in order to produce narrower particle-size distributions: (1) the energy transfer from gas to liquid should be uniform; (2) the

TABLE I

Selected References

| | Dependent Variable | | |
Independent Variable	Average Particle Size	Particle-Size Distribution	Particle Shape
Atomizing Fluid			
Intrinsic density (type of gas)	10	10	
Velocity, density	1, 2, 4, 5, 8 11, 12, 13	1, 5, 6, 8 11, 12	
Temperature	1, 3	3	
Viscosity	10	10	
Liquid to be Atomized			
Viscosity	2, 13		
Surface tension	2		
Specific gravity	2		
Velocity	13		
Composition			7
Nozzle Design			
Geometry of fluid flow	4, 11, 13, 14	1, 4, 11	
Mass flow rate of fluids	1, 2, 4, 5 11, 12	2, 4, 5, 6, 12	
Diameter of metal stream	1, 2, 3, 7, 8, 9, 13	1, 2, 3, 7, 8, 9	
Median particle size		2, 3, 4	

energy transfer from gas to liquid should be as efficient as possible. Nozzle designs conforming with these requirements will be discussed in the following section. Table I contains a selection of references dealing with the effect of nozzle design and metal and atomizing fluid properties upon average particle size, particle-size distribution, and particle shape. All practically useful correlations reported are of an empirical nature. Information is particularly scarce in the case of particle-size distribution. Several authors have shown that $\sigma_g = ad_m^b$ (a, b = constants), i.e., a plot on double logarithmic paper of the geometric standard deviation, σ_g, versus the mass median diameter, d_m (in our case equal to the geometric mean diameter, d_{gm}), produces a straight line [2, 4].

Nozzle Designs

On the basis of uniformity and efficiency of energy transfer as well as operational ease, we can distinguish the two types of twin-fluid atomization of metals

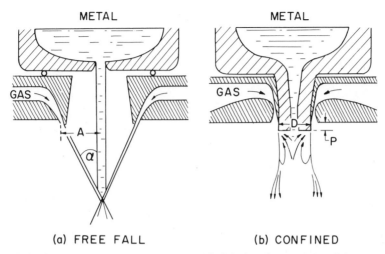

(a) FREE FALL   (b) CONFINED

Figure 3. Two-fluid atomization designs. Critical design characteristics: [a] (1) angle α; (2) distance A; (3) annular slot vs discrete jets; (4) tangential vs axial inlet of gas. [b] (1) angle α; (2) protrusion P; (3) diameter of metal nozzle D; (4) tangential vs axial inlet of gas.

shown schematically in Figure 3. In design (a), the so-called "free-fall" design [2,8,12], the metal is permitted to travel unconfined from the exit of the tundish to the point of interaction with the gas. All designs for the water atomization of metals, and many designs for the gas atomization of metals, are of this less complex type. From an operational and technical point of view, the most important design characteristics of the free-fall type of nozzle are the angle formed between metal and gas stream, the distance between the orifice of the gas nozzle(s) and the axis of the metal stream, the use of a continuous annular slot or of a plurality of discrete jets, and the tangential component (if any) of the gas stream. Characteristically, all of these critical design characteristics display interaction effects; and certain combinations of these variables can lead to unsatisfactory operation. Nozzle fouling, for instance, occurs when the metal stream contacts and adheres to the inner cylindrical wall of the atomization head, either due to an instability of the metal stream or due to back-pressure. This type of nozzle fouling can be eliminated by increasing the distance between the metal stream and the orifice of the gas nozzle.

In design (b), the so-called "confined" design [1,4,12,15,16], metal disintegration starts at the exit of the nozzle. This maximizes the uniformity of energy transfer from gas to metal. Further, as indicated by the arrows of Figure 3(b), this design effects a prefilming of the liquid, *i.e.*, the liquid forms itself into the shape of a hollow cone, the outer diameter of the hollow cone being determined by the diameter of the nozzle. The prefilming feature is demonstrated in Figure 4, where water was used instead of liquid metal. The vacuum generated by the flowing gas pulls the water from the tube and carries it, against gravity, to

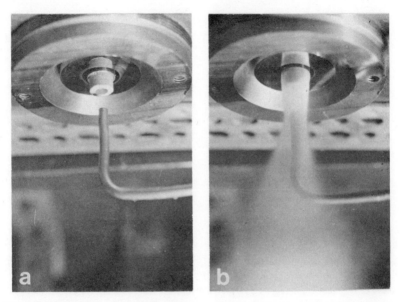

Figure 4. Illustration of prefilming feature of twin-fluid atomization design of Figure 3(b): (a) nozzle design plus tube connected to water reservoir; (b) same but with gas flow On.

Figure 5. Examples of metal build-ups obtained with nozzle design of Figure 3(b).

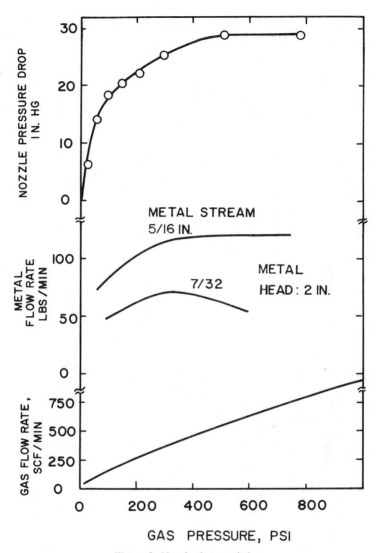

Figure 6. Nozzle characteristics.

the edge of the nozzle as shown in Figure 4(b). The flow pattern of the atomized water appears identical to the pattern that results when the water issues from the nozzle as in regular atomization.

From an operational point of view, the confined design is much more design-sensitive than the free-fall design. The nozzle for the liquid metal must reasonably withstand both the corrosion and erosion from the liquid metal and the high-velocity gas, as well as the thermal shock generated by the simultaneous impingement of the fluid media. Figure 5 shows examples of metal build-up at the periphery of the nozzle. Whenever such build-ups form, they usually form at the start of atomization and grow to unacceptably large sizes within less than a minute. They may be the result of an unfavorable combination of critical design characteristics or of excessive heat abstraction at the edge of the nozzle [1,12].

Atomization Data on Prefilmed Cu and Cu–0.33% Al

The data reported in this section were obtained with the nozzle design shown in Figure 3(b). ($P = \frac{3}{8}''$, $\alpha = 0°$, $D = 0.5''$, tangential gas inlet). Melting was done in clay-graphite crucibles under a protective atmosphere of hydrogen. The pouring temperature was kept constant at 2350°F.

Figure 6 shows some nozzle characteristics. Note that the metal flow rate is a function of the gas pressure, and that for small metal orifices the flow rate exhibits a maximum.

Figure 7 shows the effect of gas-to-metal mass ratio on the mass median diameter of a Cu–0.33% Al alloy. For a given mass-median diameter, the gas consumption is significantly smaller for the larger metal stream. This relationship also holds when the mass median diameter is related with the power requirements

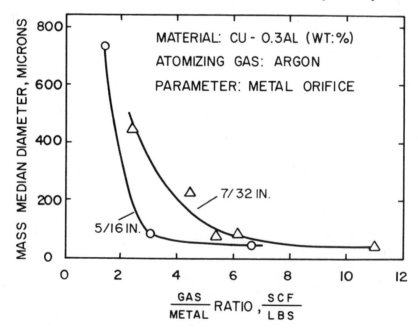

Figure 7. Effect of gas/metal ratio on mass median diameter.

TABLE II

Effect of Diameter of Metal Stream and Geometric Mean Diameter of
Argon-Atomized Powder on Geometric Standard Deviation and Power Requirements

Diameter of Metal Stream, in.	d_{gm}, microns	σ_g	Power Requirements*	
			$\dfrac{\text{erg} \times 10^{10}}{\text{lb metal}}$	$\dfrac{\text{erg} \times 10^{7}}{\text{cm}^2 \text{ metal}}$
$\frac{7}{32}$	445	3.97	9.4	3.0
	82	2.32	58	6.1
	44	1.98	128	8.2
$\frac{5}{16}$	730	4.90	5.6	2.1
	80	2.56	29	2.7
	48	2.19	83	5.4

*Calculated on the basis of isothermal expansion of the gas.

per unit weight or per unit surface area of metal (see Table II). The geometric standard deviation, σ_g, however, increases with increasing diameter of the metal stream. An even more pronounced increase of σ_g results when pure copper is taken instead of the Cu–0.33% Al alloy (Figure 8). The σ_g of a powder is probably affected by the degree of prefilming, which in turn is affected by the surface tension and viscosity of the liquid metal.

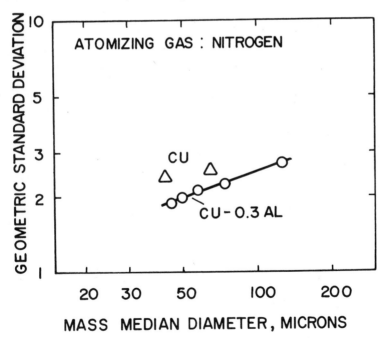

Figure 8. Effect of composition on geometric standard deviation.

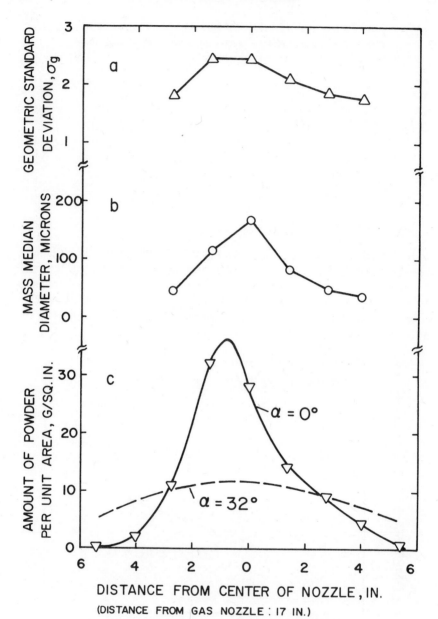

Figure 9. Effect of patternation on d_{gm} and σ_g.

Figure 9 shows data on the spatial distribution of the powder ("patternation") during atomization. In the vicinity of the axis of the metal stream there exists a pronounced maximum. Patternation can be controlled by the angle α (Figure 3) formed between gas and metal stream. The maxima of the geometric standard deviation, σ_g (curve (a), Figure 9), and the mass median diameter (curve (b), Figure 9) are directly related to the patternation (curve (c), Figure 9), and explained by coalescence and agglomeration of powder particles. Powder samples taken from areas of low concentration consist largely of individual particles, while samples taken from areas of high concentration contain many agglomerated and coalesced particles.

Summary

Selected aspects of the gas atomization of metals have been discussed with major emphasis on items of particular interest to the powder producer. These include the following: methods of atomization; such powder characteristics as mass mean particle diameter (d_{gm}), particle-size distribution (σ_g), particle shape and oxygen content; powder yield or efficiency of atomization and factors affecting them; and nozzle design.

The twin-fluid method of atomization with external mixing dominates all others for industrial use in atomization of metals. Both free-fall and confined nozzle designs are used. The latter, although more sensitive to design, is most efficient.

The mass mean particle diameter (d_{gm}) and particle-size distribution (σ_g) are important powder characteristics from the standpoint of production costs. Atomizing conditions affecting these are pressure of atomizing medium, ratio of metal to atomizing medium flow rates, metal composition, diameter of metal stream, nozzle design, and metal temperature. In general, particle size decreases with increase in pressure of atomizing medium, decrease in the ratio of metal to atomizing medium flow rates, and with increase in metal temperature. σ_g varies with the metal being atomized and the diameter of the metal stream.

Acknowledgment

The authors wish to express their appreciation to A. E. Petrosh and N. Phillips for their assistance with the experimental portion of this work and to the management of Metals Group, Glidden-Durkee Division of SCM Corporation, for permission to present this paper for publication.

References

1. Thompson, J.S., "A Study of the Process Variables in the Production of Aluminum Powder by Atomization," *J. Inst. Metals,* 74 (1948), 101.
2. Lubanska, H., "Correlation of Spray Ring Data for Gas Atomization of Liquid Metals," *J. Metals,* 22, no. 2 (1970), 45.
3. Tamura, K. and Takeda, J., "A Study on Production of Copper Powder by Atomization," *Trans. Nat. Res. Inst. Met.,* 5 (1963), 252.
4. Gretzinger, J. and Marshall, W.R., "Characteristics of Pneumatic Atomization," *AIChE J.,* 7 (1961), 312.
5. Bitron, M.D., "Atomization of Liquids by Supersonic Air Jets," *Ind. Eng. Chem.,* 47 (1961), 23.
6. Dunskii, V.F., "On Coagulation During the Atomization of a Liquid," *Sov. Phys.-Tech. Phys.,* 1 (1957), 1232.
7. Naida, Yu. I., and Nichiporenko, O.S., "Atomization of Liquid Bronze and Properties of the Resulting Powders," *Sov. Powder Met. Metal Ceram.,* no. 7, 55 (1967), 525. Trans. of *Poroshkov. Met.,* no. 7, 55 (1967), 23.
8. Nichiporenko, O.S., "Investigation of the Process of Bronze Atomization by Air," *Sov. Powder Met. Metal Ceram.,* no. 12, 60 (1967), 947. Trans. of *Poroshkov. Met.,* no. 12, 60 (1967), 10.
9. Nichiporenko, O.S., Naida, Yu. I., and Kochergin, A.V., "Production of Nickel Powder by Spraying," *Poroshkov. Met.,* no. 12, 96 (1970), 1 (in Russian).
10. Lewis, H.C., Edwards, D.G., Goglia, M.J., Rice, R.I. and Smith, L.W., "Atomization of Liquids in High Velocity Gas Streams," *Ind. Eng. Chem.,* 40 (1948), 67.
11. Fraser, R.P., Eisenklam, P. and Dombrowski, N., "Liquid Atomization in Chemical Engineering: Part 4–Twin-Fluid Atomisers," *Brit. Chem. Eng.,* 2 (1957), 610.
12. Schellenberg, R., "Zerstäubung von Metallen, insbesondere von Zink," Unpublished Ph.D. dissertation, Technische Hochschule Stuttgart, 1964.
13. Weiss, M.A. and Norsham, C.H., "Atomization in High Velocity Air Streams," *Ars. J.,* 29 (1959), 252.
14. Priem, R.J., "Breakup of Water Drops and Sprays with a Shock Wave," *Jet Propul.,* 27 (1957), 1084.
15. U.S. Patent No. 1,501,449, July 15, 1924, "Metal Disintegrating Apparatus," E.J. Hall to Metals Disintegrating Company, New York. U.S. Patent No. 1,545,253, July 7, 1925, "Nozzle Intended for Use in Disintegrating Apparatus," E.J. Hall to Metals Disintegrating Company, New York.
16. U.S. Patent No. 3,253,783, May 31, 1966, "Atomizing Nozzle," R.L. Probst, C.H. Sayre, and P.I. Karp to Federal-Mogul-Bower Bearings, Inc., Detroit, Michigan.

4. Coarse Powder Techniques

R. WIDMER
Industrial Materials Technology, Inc.
Woburn, Massachusetts

ABSTRACT

Since powders processed by hot consolidation do not necessarily have to be fine- and cold-moldable, this opens up a new field of particle metallurgy, different from classical powder metallurgy. Experiments were conducted with coarse powders of iron-, nickel-, and cobalt-base alloys as well as some titanium alloys. Coarse powders were consolidated by isostatic hot pressing and hot extrusion. Attractive mechanical properties were obtained.

Coarse powders offer some advantages over fine powders primarily because contamination is less of a problem. In alloys with high-temperature creep requirements, microstructures with a large grain size can be developed rather easily.

Introduction

The potential advantages of powder processing over conventional casting and ingot-conversion techniques are well known. They are mostly based on the more rapid cooling rate observed in a small metal particle as compared to that of an ingot. The solidification of a large piece of metal leads to a variety of structural flaws, both on a macro and a micro scale. It is the uniformity and homogeneity of the microstructure that give a powder product a much better chance for reproducible properties and, at the same time, better mechanical properties.

Powder metallurgy has a reasonably long history within the metals industry; however, the applications of the past were mostly confined either to highly specialized products such as molybdenum and tungsten, or alternatively to powders used for the fabrication of low-cost cold pressed and sintered parts. Powder parts of the latter variety have found increasing application because they allow for close control of tolerances, low scrap losses, applicability to complex shapes, and rather good reproducibility. Since conventional processing includes a cold pressing step, compactibility at room temperature was an important requirement for powders. Among other things, this called for a rather small particle size and, preferably, irregular shapes. The lack of full density of pressed and sintered products restricted the applicability of powder metallurgy in the broadest sense. In particular, it excluded most complex high-temperature alloys and high-speed steels.

With the advent of improved atomization techniques for the fabrication of complex alloy powders, as well as improvements and innovations in hot consolidation, the ground rules were changed. By using hot working methods such as extrusion, isostatic hot pressing, and direct powder rolling it is possible to fully densify a powder without the use of a liquid phase. The densification process is not primarily dependent on diffusion but rather on high-temperature plastic deformation of the particles.

It is for these reasons that the formerly imposed restrictions on particle size and shape do not hold any more. The problem for the metallurgist now is to determine the optimum particle size and shape of a given material to be consolidated to a given end-product and, in addition, to define the best method of consolidation.

Coarse particles have several advantages over fine powders: they are usually cheaper to fabricate; they do not contaminate as easily and, therefore, do not offer storage problems; and they can generally be more easily handled.

Prior to a discussion of some recent experiments with coarse powders of various alloys, it is of interest to point out some industrial developments in which large metal particles played the key role.

In the Reynolds Metal particle sheet program, coarse aluminum particles are produced by a centrifugal atomizing technique. These particles are directly hot rolled to sheet. Whereas oxides formed on each particle prevent aggregation prior to rolling, they do not prevent pressure welding during the actual rolling operation. The process can be applied in air for aluminum, but other metals would lend themselves equally well to direct powder rolling, provided a protective atmosphere is used. Although technical feasibility of coarse powder rolling to steel sheet and strip was demonstrated successfully (for example, by BISRA in Great Britain), and the economics of the process looks attractive, no important inroads were made so far in this area. This in part is probably due to the reluctance of steel companies to replace existing rolling equipment. The situation would most likely be quite different if a new plant were to be built.

It is the purpose of this report to show with a few examples of superalloys and other high-strength materials, what types of coarse powders can be obtained with present technology, how these powders can be hot consolidated, and what microstructures and mechanical properties can be expected.

Results and Discussion

Most of the examples discussed here are steam-atomized powders subsequently surface cleaned and hot consolidated by isostatic hot pressing, extrusion, or both in sequence. One material—a titanium alloy powder—was obtained by an inert processing technique. In choosing a method of coarse powder manufacture one had to trade, to some degree, between effective and rapid cooling on the one side, and particle morphology and surface cleanliness on the other. In several alloy systems it was found that steam atomization followed by chemical surface cleaning can yield a good-quality coarse powder, allowing the fabrication of billets or bar stock with uniform microstructure and good mechanical properties. This observation applies, for example, for many low-alloy steels, stainless steels, cobalt-base alloys and non–age hardening nickel-base alloys. For materials containing an appreciable amount of reactive metals, complete inert gas or vacuum atomization is probably the only feasible processing route. From the point of view of the processing costs involved, steam atomization looks extremely attractive, both with regard to cost of operation and yield of usable powder. The added cost of surface cleaning is negligible.

Powder Manufacture and Characterization

Coarse powders with a particle size range of 0.5–5 mm were produced by steam atomization followed by water quenching. Cooling rates were in the range of 10^2–10^3 °C/sec, with resulting secondary dendrite arm spacings of 5–10 microns. The change in the microstructure from a cast bar to a coarse powder is illustrated in Figures 1 and 2 with the nickel-base alloy IN-100. Dendrite arm spacings are about the same as those found in fine argon-atomized powder. Whereas very close composition control was maintained with any of the iron-, nickel-, and cobalt-base alloys during melting, atomization caused losses of reactive elements connected with surface oxidation. The maintenance of the desired particle size range was no problem but the shapes of the particles varied widely, dependent on the composition in addition to conditions of atomization. Figures 3–7 give an illustration of the varying particle morphology encountered in the range of alloys atomized. Maraging steel, for example, yields practically perfect spheres, whereas Hastelloy X has irregularly shaped but compact particles. Cobalt-base alloys such as X-45 and MAR-M 509 atomize into reasonably round but often hollow spheres. Udimet 710, an age-hardening nickel-base alloy, emerges in the form of twisted flakes, presumably because the high-melting-point aluminum and titanium oxide skins "cage" the particles in their initial distorted liquid shapes and prevent surface-tension forces from rounding the liquid metal droplets. A significant measure of the flakiness of this powder is the very low packing density of 0.3, versus about 0.5 for the other materials.

Oxygen contents of all powders were quite high in the as-atomized condition, mostly due to surface oxidation, and it was therefore mandatory to include a surface cleaning step. Since most powders were not ideally round, simple

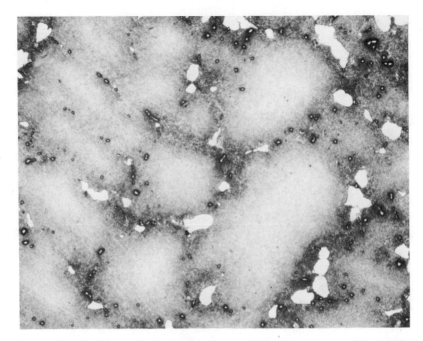

Figure 1. IN-100. Three-inch-diameter cast bar. Solidification structure. Etched, 200X.

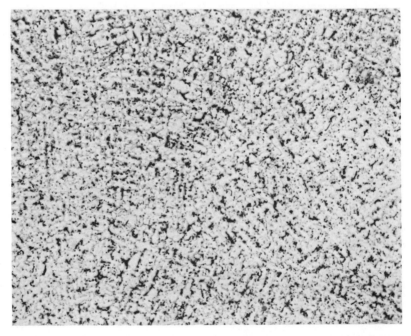

Figure 2. IN-100.   Powder particle dendrite structure.   Heat No. 149, $-3\frac{1}{2}$/+4 mesh. Etched, 200X.

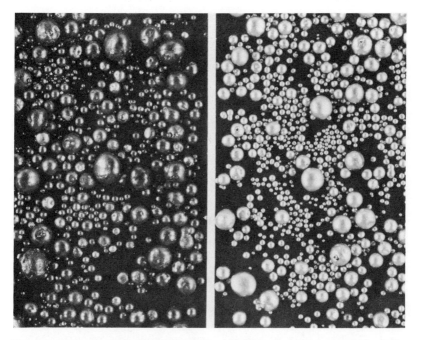

Figure 3. Vascomax 300. Powder sample, $-4/+30$ mesh. Steam-atomized. Left, as-atomized; right, surface cleaned. 1X.

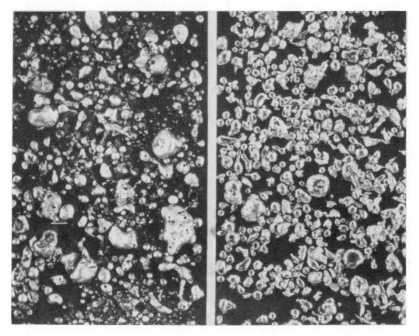

Figure 4. Cobalt-base alloy X-45. Powder sample $-4/+30$ mesh. Steam-atomized. Left, as-atomized; right, surface cleaned. 1X.

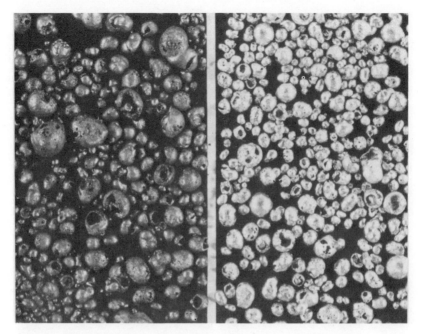

Figure 5. Cobalt-base alloy MAR-M 509.   Powder sample ⁻4/+30 mesh.   Left, as-atomized; right, surface cleaned. 1X.

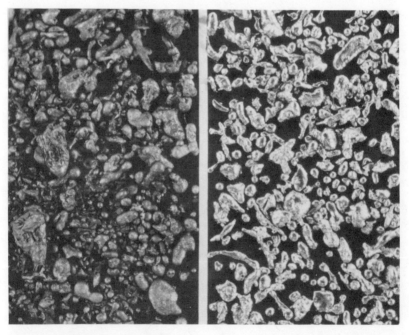

Figure 6. Nickel-base alloy Hastelloy X.   Powder sample ⁻4/+30 mesh.   Left, as-atomized; right, surface cleaned. 1X.

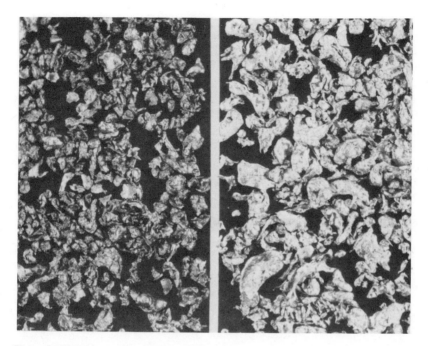

Figure 7. Nickel-base alloy Udimet 710.   Powder sample −4/+30 mesh.   Left, as-atomized; right, surface cleaned. 1X.

mechanical cleaning was not sufficient and an additional chemical cleaning step had to be included. Both salt-bath and wet-chemical methods were applied. This additional powder processing step brought the oxygen levels down to reasonable levels, sometimes as low as that of the remelt stock, as in the case of Hastelloy X. (See Table I.)

Consolidation of Coarse Powders

For consolidation hot isostatic pressing, extrusion and hot isostatic pressing followed by extrusion were used. In all instances, powders were packed into

TABLE I

Oxygen Content of Various Surface-Cleaned Coarse Powders

Alloy Designation	Oxygen Content (ppm)	For Comparison: Oxygen Content of Melt Stock (ppm)
Hastelloy X	320	314
X-45	180	109
Udimet 710	210	30

mild steel cans without prior cold compaction. Cans were evacuated to 10^{-5} mm Hg up to a temperature of 900°F prior to sealing.

Depending on the alloy to be processed, hot isostatic pressing was carried out with dwell times of 1–3 hours, at a temperature of 2000–2300°F and a pressure of 15,000–30,000 psi. In no case was a liquid metal temperature reached.

Extruded bar stock was produced directly from canned coarse powders using a 16X extrusion reduction at 2100–2200°F.

Both methods of consolidation yielded fully dense material under optimized processing conditions. Whereas extrusion is always limited to bar stock and relatively simple profiles, isostatic hot pressing can yield rather complex finished shapes as well as large-diameter billets weighing several tons.

Microstructure and Mechanical Properties

The microstructure of some selected alloys was studied and some typical examples are included in Figures 8–14. Coarse powders consolidated by either method were found to be fully dense, with a uniform distribution of second-phase particles. The grain size of the isostatically hot pressed samples was either about the same as the original particle size or, more often, somewhat smaller. Extruded samples had a uniform fine-grain size.

Figure 8. Microstructure of Inco 713LC isostatically hot pressed fine powder. As-pressed. 100X.

It is of interest to compare fine and coarse powder compacts in their heat-treatment response with particular attention to grain growth. Although advances were made in the atomization of fine powders it is still generally very difficult, if not impossible, to grow large grains by thermal treatment only. Figure 8 shows the microstructure of an isostatically hot pressed Inco 713 LC fine-powder sample, which was heat-treated after pressing at various temperatures and times up to 2300°F. The grain size did not change but stayed around the same as the original particle size.

Figures 9 and 10, on the other hand, illustrate the heat-treatment response of an extruded bar made from a coarse powder of X-45. A rather small grain size can be observed in the extruded condition, but it can be grown into a size much larger with a subsequent heat treatment.

Particle boundaries, or rather areas of former particle boundaries, have a very special importance in considering high-temperature performance. For good high-temperature creep strength one would want a large grain size and, therefore, a minimum amount of grain-boundary area. In addition to the grain size, one has to consider the structure of the grain boundaries, since it is not identical in cast, wrought, and powder materials.

As a representative of a high-strength cast cobalt-base alloy, MAR-M 509 was processed via the coarse-powder route and isostatic hot pressing. The microstructure of the final product is illustrated in Figures 11 and 12. The first of the

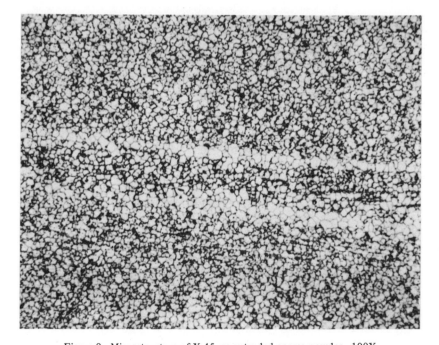

Figure 9. Microstructure of X-45, as-extruded coarse powder. 100X.

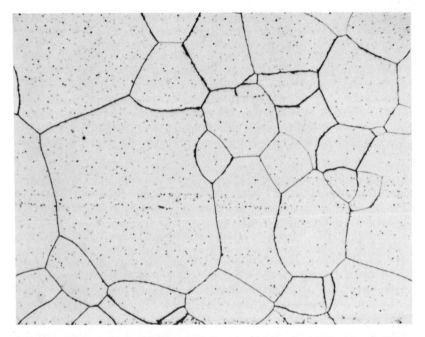

Figure 10. Microstructure of X-45 extruded coarse powder after heat treatment; 16 hours at 2375°F. 100X.

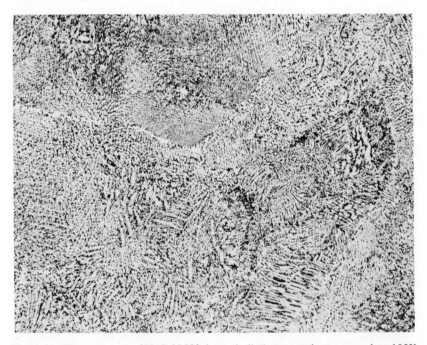

Figure 11. Microstructure of MAR-M 509, isostatically hot pressed coarse powder. 100X.

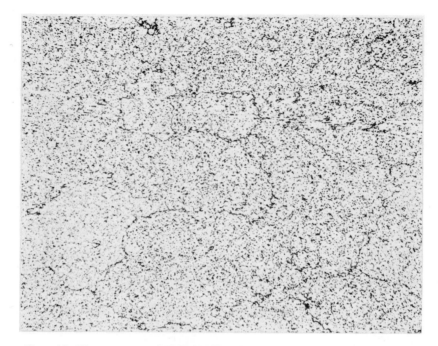

Figure 12. Microstructure of MAR-M 509. Coarse powder isostatically hot pressed, extruded, and heat-treated 4 hours at 2250°F. 100X.

two photomicrographs shows the material in the as-pressed condition. Although there is still some indication of a dendritic structure, the carbide particles are quite finely dispersed—quite different from a regular cast material. After a one-step high-temperature heat treatment, some of the carbides dissolve and boundaries of a large-grained microstructure become apparent. It is of interest to note that no appreciable carbide-particle growth has taken place.

A continuing puzzle is the topic of particle boundaries, or the areas of former particle boundaries, in the consolidated structure. Mostly, there does not seem to be any lack of bonding, and yet quite frequently the former powder particles are clearly delineated and, as it turns out, these boundaries often represent areas of some weakness.

A case is illustrated in Figures 13 and 14, which show the microstructure of a high-speed steel powder product. In the low-magnification photomicrograph, one can easily recognize the boundaries of the spherical particles. If those areas are scrutinized more closely at a higher magnification, one can observe a higher concentration of carbide particles than present in the interior of those spheres. No lack of bonding can be found, but this redistribution of second-phase particles can certainly have an effect on some mechanical properties. Similar observations were made in some nickel-base superalloys, particularly those containing more than 0.05 percent carbon.

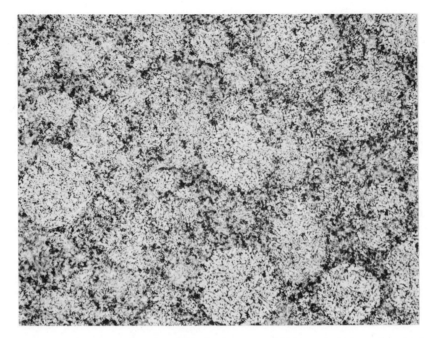

Figure 13. Microstructure of isostatically hot pressed high-speed steel powder sample. As-pressed. 100X.

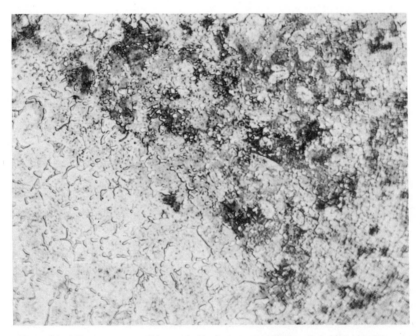

Figure 14. Microstructure of isostatically hot pressed high-speed steel powder sample. As-pressed. 1000X.

It is of interest to consider some of the mechanical properties obtained from extruded or isostatically hot pressed coarse powders. Figure 15 represents a parametric plot with stress-rupture properties of Hastelloy X and X-45. The two straight lines represent typical values found in handbooks for cast X-45 and wrought bar stock of Hastelloy X. The two sets of individual points are stress-rupture data points obtained on coarse-powder extrusion of the materials at 1500 and 1800°F. Both materials had been heat-treated to a large grain size, as illustrated in Figure 10. Alloy X-45 shows values equal to those of cast material at both temperatures, whereas the Hastelloy-X points drop off slightly with the higher temperature. The point demonstrated here is the effect of grain size on high-temperature creep. It is also conceivable that it is not the actual grain size only that has an influence on the rupture life, but that possibly the total particle-boundary area is of importance.

Other mechanical properties illustrate observations generally made in P/M materials. With a fine grain size, one gets very high low-temperature strength, poor high-temperature strength—all coupled with high ductility values. With larger grain size, the high-temperature strength is improved. For typical room-temperature tensile data, see Table II.

In a further example, powder of the titanium alloy with 6 percent aluminum, 6 percent vanadium, and 2 percent tin was consolidated by isostatic hot pressing. In an attempt to optimize processing parameters, powder samples were pressed at 1200, 1300, and 1500°F. Representative photomicrographs are shown in Figures 16–19. The lowest temperature is obviously too low for complete con-

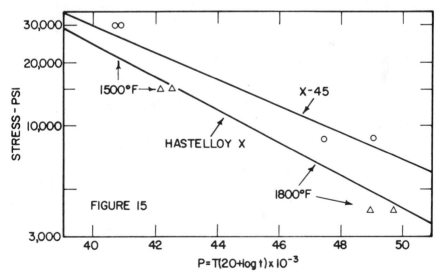

Figure 15. Stress-rupture properties of extruded and heat-treated bar stock of "clean" coarse powders.

TABLE II

Typical Room-Temperature Tensile Properties of Coarse Powder Alloys

Alloy	UTS (psi)	0.2% Yield (psi)	ϵ (%)	RA (%)
Hastelloy X	114,000	52,100	26	25
X-45	163,000	84,000	28	24
U-710	157,000	146,000	4	7
MAR-M 509	195,000	123,000	18	14

solidation, although a longer dwell time might achieve the closing of the remaining holes and voids. Also, the individual particles still have their dendritic structure. At 1300°F, nearly 100 percent density is achieved, with the exception of some individual small holes. At the highest processing temperature, full density was achieved. Second-phase particles are redistributed into a very fine pattern.

This series of experiments shows that a powder not compactible at room temperature can rather easily be fully consolidated at a moderate temperature, as long as processing conditions of easy plastic deformation exist. In this, isostatic

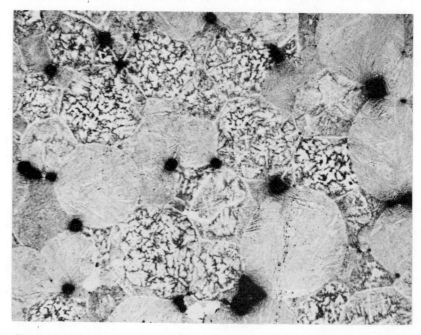

Figure 16. Microstructure of titanium 6Al–6V–2Sn alloy isostatically hot pressed at 1200°F. As-pressed. 100X.

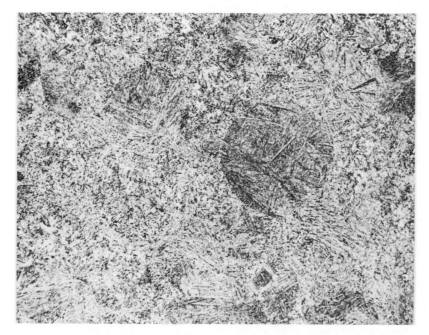

Figure 17. Microstructure of titanium 6A1–6V–2Sn alloy isostatically hot pressed at 1300°F. As-pressed. 100X.

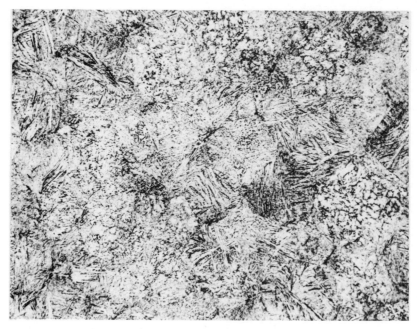

Figure 18. Microstructure of titanium 6A1–6V–2Sn alloy isostatically hot pressed at 1500°F. As-pressed. 100X.

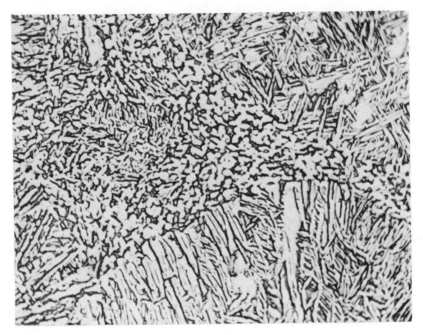

Figure 19. Microstructure of titanium 6Al–6V–2Sn alloy isostatically hot pressed at 1500°F. As-pressed. 1000X.

hot pressing turns out to be an effective tool, since it allows for the variation of temperature, pressure, and time.

Summary and Conclusions

Experiments with coarse powders of various materials showed that a wide range of particle sizes can be considered for hot consolidation, such as extrusion or isostatic hot pressing. The inclusion of larger particles—up to 5 mm in diameter—for powder processing opens up some attractive possibilities, without a sacrifice of the usual advantages, such as homogeneity and uniformity of microstructure. Coarse powders are easier to handle than fine ones, and usually cheaper to fabricate. Large grains can readily be grown in high-temperature alloys made from coarse powders, thus enhancing the high-temperature creep strength.

5. Specialty Methods of Powder Atomization

N. J. GRANT

Massachusetts Institute of Technology
Cambridge, Massachusetts

ABSTRACT

Powder-metallurgy techniques for parts production are now being extensively supplemented by powder metallurgy for production of semifinished and finished fully dense alloys by hot mechanical and hot isostatic pressing and by hot working processes such as rolling, forging, and extrusion. Of particular interest are the more highly alloyed materials which permit large savings in yield of useful product, machining costs, and the attainment of superior mechanical and physical properties. Materials of much higher alloy content and higher cost have fostered a variety of new atomization techniques and other powder-producing methods. These methods are worthy of examination because of their potential for commercial production of powders in conventional as well as exotic materials.

Introduction

Little or no reference will be made to any of the current production techniques such as high-pressure water atomization, high-pressure inert gas atomization, sponge-iron production, the Sherritt Gordon precipitation process, electrolytic techniques, hydrogen reduction, or ball-milling of coarser aggregates. In fact, some of these are well covered in other chapters in this publication.

The promise and potential of a number of emerging processes are instead of interest in this chapter because of special features which open new doors to powder-metallurgy processes and products which are of commercial promise.

Powder Characterization

The literature is abundantly supplied with methods of powder characterization, ranging through powder size, shape, size distribution, surface area, analysis, etc.

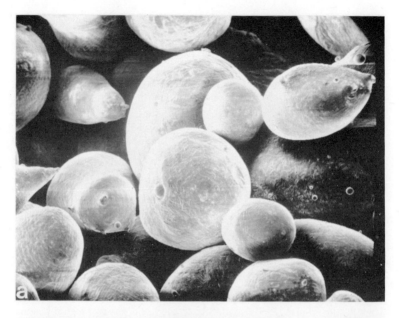

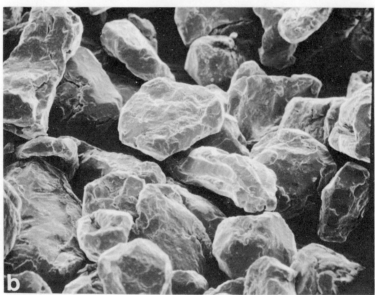

Figure 1. SEM views of powders produced by three techniques. (a) Inco 713C, inert gas-atomized, 240X. (b) Electrolytic iron, crushed, 200X. (c) Carbonyl nickel, as-deposited, 2000X.

In a repetitive process, such as in iron-sponge manufacture, these and other measurements are meaningful for purposes of control, comparison, or classification. When different processes of powder production are compared, however, many of these same measurements only lead to confusion and misinformation. For example, what is the significance of a powder size and shape measurement when comparing an electrolytic (crushed) powder, an inert gas-atomized powder, and a carbonyl product? Figure 1 compares scanning electron micrograph views of representative powder particles produced by these three methods. There are unbelievably large differences in the surfaces which one does not usually see and, therefore, differences in surface area, reactivity, degree of contamination (or difficulty in cleaning), etc.

Wherever possible, scanning electron micrograph views will be used to illustrate the powders produced by the specialty powder-making technique, at magnifications which are in themselves almost fully descriptive.

Specialty Techniques

No effort is made to present these techniques in any particular order, because the relative importance or potential importance among them is not known today.

1. *The Rotating-Electrode Process (REP).* Whittaker Nuclear Metals Division produces high-quality powders of low oxygen content by rotating a rod, the front end of which is heated by an electric arc or a plasma, as shown in Figure 2.

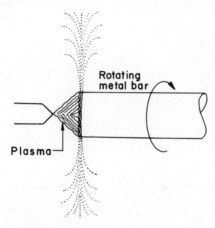

Figure 2. Rotating-electrode process of powder production.

Molten droplets are thrown by centrifugal force and collected in an inert gas-filled chamber [1]. Helium is used to try to achieve a somewhat more rapid quench of the particle, but the effect is undoubtedly small. The powder size range is relatively narrow, and neither fine nor very coarse powders are produced. The range of powder sizes falls between about 30 and 500 microns. The actual particles are quite spherical, expose a smooth surface of relatively small area (see Figure 3), and are free of major porosity and surface defects.

Oxygen contents of REP powders show 75 ppm for IN-100 and less than 30 ppm for Maraging 300, both excellent values.

Maraging 300 steel produced by hot isostatic pressing (HIP) plus hot extrusion of REP powders, in the aged condition (3 hours at 900°F) gave a yield strength of 285,000 psi, an ultimate of 291,000 psi, and 47 percent reduction of area (compared to 270,000 psi, 280,000 psi, and 48 percent, respectively, for commercial ingot product). No banding was observed and the grain size was uniformly fine.

2. *Soluble Gas Atomization (Hydrogen Evolution Atomization).* In stream degassing of large steel ingot castings, wherein the molten steel is poured through a (fair) vacuum, it was observed that even relatively small amounts of dissolved hydrogen (as little as 1–3 ppm) resulted in disintegration of the metal stream as hydrogen (and some CO) were precipitated. A similar approach has been utilized by Homogeneous Metals Company to atomize liquid alloys, especially some of the more highly alloyed materials such as IN-100, an alloy which requires either a vacuum or an excellent inert atmosphere to produce useful clean powders.

Figure 3. IN-100 powder produced by the rotating-electrode process. 253X.

Very low values of oxygen are reported. On hot isostatic pressing, an apparent dense structure results (at 30,000 psi and 2000–2300°F), with some finely scattered porosity. Clearly HIP is able to confine the porosity due to the presence of unpurged hydrogen; however, reheating to very high temperatures permits expansion of the H_2 gas. Obviously, vacuum treatment of these powders is necessary.

The use of vacuum leads to production of clean powders; powder shape features are not controllable by this technique. Powders smaller than about 100 microns tend to be spherical, whereas coarser powders are primarily flake-shaped. Yield of powders on a given screen are poorly reproducible. Excess hydrogen must be removed; cost features for larger scale production are not known; screen analyses and yield of powder in various size fractions are not known for any given alloy.

3. *Ultrasonic Disintegration.* Liquid metals and alloys have low shear resistance (fluidity is very high). High-velocity fluids (water, steam, air, inert gases) also readily disintegrate liquid metals; atomization is essentially assured. What are not guaranteed in high-velocity fluid atomization are particle shape, size, size and shape distribution, yield of powder in each size range, etc.

Theoretically, because of the sharply defined shearing forces—at much higher frequency than is possible in high-pressure, high-velocity atomization processes—ultrasonic atomization should produce narrower size fractions, more repetitive powder shapes, and much finer powders on average.

With the low-melting metals and alloys (or other low-temperature fluids), a large number of clever ultrasonic devices have been developed and used successfully to produce fine near-micron mists, powders, etc. An ultrasonically vibrating, freely suspended beam readily produces fine metallic powders from a liquid stream poured on to a vibrating end. Erosion of the beam material can and does take place; if oxidation is also a problem, then the atomization of higher melting metals and alloys is difficult and costly. Metallic, ceramic, and intermetallic beam materials have been tried.

In Sweden [2], however, an ultrasonic die (generator) was designed which involved no direct contact between the generator and the liquid metal stream. Capable of operating in the 60,000–120,000 cps range, liquid metals are disintegrated to a particle size range from less than 1 micron to perhaps 50 microns (with up to 75–90 percent in the −10 microns size range).

Basically, a typical generator is a small circular die about 1″ tall, perhaps 1.5–2″ OD, and 1/2–3/4″ ID. A circular manifold near the top outer portion feeds high-pressure gas at 15–30 atmospheres pressure into 20 or 30 precision ducts which fire pulses of gas at a characteristic frequency onto the metal stream being poured through the ID of the generator. The stream is finely atomized, producing near-spherical powders. Figure 4 shows a characteristic shadowgraph (at 1000X) of stainless steel powder produced by this technique, yielding an average particle size of about 4 microns.

Since inert gases can be used for atomization, clean, fine powders are possible. The relative fineness should permit rapid sintering and, perhaps, easy cold or hot compaction. Alloy composition is not a limitation, although the fine powders have to be protected during subsequent exposure to the atmosphere.

Compacted and extruded bars of the 18-8 alloy (which contained several percent of Cr_2O_3 as a fine dispersion) showed a yield strength of 85,000 psi and an ultimate tensile strength of 112,000 psi. The ductility was relatively poor due to the presence of the chromium oxide.

4. *Centrifugal Atomization.* For practical reasons, centrifugal atomization is restricted to the lower melting point metals and alloys; the wear, solution, oxidation (corrosion), and creep of the centrifugal spinners restricts the temperature which may be used. Metals such as lead, tin, and aluminum have been centrifugally atomized. In the case of lead, even fine ribbons with aspect ratios of 30 or 50 to 1 have been reported; lower aspect ratios, tending toward a spherical powder, are easier to achieve.

The Reynolds Metals Company [3] has produced acicular and tear-drop shaped powders by continuously atomizing liquid aluminum alloys at speeds near 3,000 rpm, using steel spinners. Yields greater than 90 percent of useful

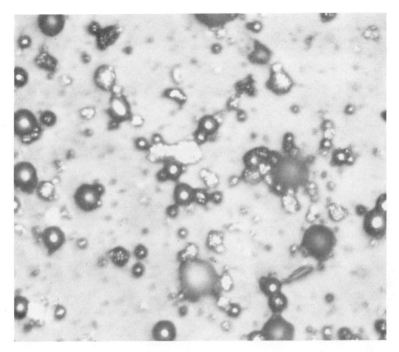

Figure 4. Fine powders of an 18-8 stainless steel produced by ultrasonic atomization. Average particle size is 4 microns; maximum particle size is about 10 microns. 1000X.

powder (0.05-0.07″ × 0.15-0.25″) have been reported. Figure 5 shows scanning electron micrograph views of aluminum alloy 2024 centrifugally air-atomized product.

The more rapid quench and resultant structural refinement of the alloy, compared to commercial ingot product, results in property improvements as shown in Table I.

Interestingly, this coarse atomized powder is fed and heated continuously through a tall furnace, then delivered to a single set of large-diameter rolls, yielding a fully dense product in coiled lengths of any reasonable desired size. Properties are marginally better than those of DC cast ingot material. The alloy exhibits significantly improved resistance to recrystallization and grain growth, and thus remains finer grained.

5. *Splat-Cooled Atomized Foils*. In the past five to seven years, rapid quenching of liquid metals in powder and flake form has advanced tremendously. The number of papers on cooling rates, powder and foil shapes, splatting techniques, structural features, structural stability, properties, etc., is now into the hundreds [5].

Basically, liquid metals and alloys are atomized into fine droplets (from less

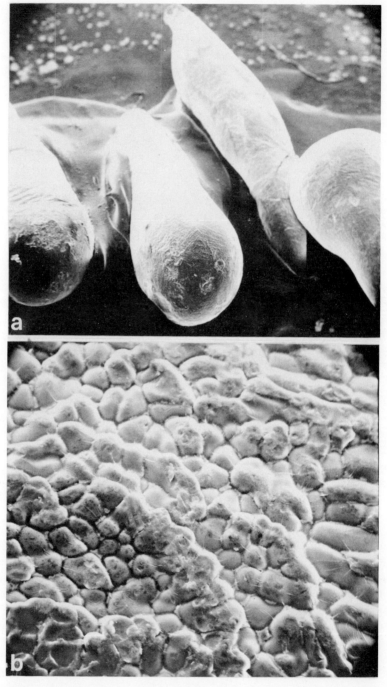

Figure 5. Scanning electron micrographs of aluminum alloy 2024 centrifugally air-atomized. View (a), acicular and tear-drop shapes, 23.5X; view (b), dendritic nature of the surface at higher magnification, 500X.

TABLE I

Tensile and Fatigue Properties of Aluminum Alloy 2024, Solution Treated and Aged as
a Function of the Cooling Rate of the Liquid Metal [4]

Alloy Source	Cooling Rate $^{\circ}$C, sec^{-1}	Y.S. (0.2%) (psi)	U.T.S. (psi)	Elong. (%)	R.A. (%)	Fatigue Life at 35,000 psi, cycles
Ingot	10^{-1}	40,200	67,200	23	37	100,000
Centrif. air atom.	10^2	42,000	70,000	26	39	300,000
Splat	10^5	47,000	76,000	22	34	700,000

Note: T–4 aged condition.

than 1 micron to perhaps 100 microns) and delivered at high velocity (while still molten) against a metallic substrate which has large quenching capacity. Cooling rates from 10^3-10^8°C/sec are attainable, leading to highly refined dendritic and grain structures; segregation is highly restricted, solute-element solubilities are enhanced to a large extent [4], and excess phase particles are highly refined and uniformly distributed within the dendrite and grain-boundary areas.

The impacted (splat) foils vary in thickness from less than 1 micron to as much as 50 microns, the length and width dimensions being of less consequence and averaging from 3–10 mm. For research purposes, foils tend to be on the low side of these figures, and would be considered difficult to handle in subsequent consolidation and working efforts. To produce splat-cooled foils of interest, coarser atomized droplets are produced which on impact against the substrate will average 20–50 microns in thickness and 5–10 mm in the other dimensions. Figure 6 shows two views of Al-7% Si splat foils which had been impacted against a rotating copper disc, in air, at 1,700 rpm. The splat foils reproduce the roughened surface of the copper disc; this roughening is done deliberately to increase the contact area, to improve the tendency of the foils to stick to the wheel, and yet permit removal from the disc to avoid build-up to splat material with subsequent decrease in quench rate.

The foils consolidate readily and present no problems in hot extrusion. The properties of aluminum alloy 2024 are shown in Table II in comparison with ingot product and a centrifugally air-atomized powder, all fully converted to dense bar product. The increase in properties is quite attractive [6].

Significantly, finer 2024 atomized liquid droplets which solidified prior to impacting the rotating copper disc showed a coarser dendritic and grain structure, due to the slower rate of cooling, and somewhat poorer properties.

By control of particle size and cooling medium, it is possible to produce spherical powders which will be cooled at rates of 10^2-10^4°C/sec. Measurements have shown that fine powders (-100 mesh; -150 microns) which are water-atomized may achieve cooling rates of 10^3-10^4, depending on material and processing variables; similar powder sizes produced by gas atomizaiton will

Figure 6. Scanning electron micrographs of Al–7% Si alloy splat-cooled rotating copper disk (–14 mesh, +30 mesh flakes). (a) is at 19X, (b) is at 48X.

TABLE II

Room-Temperature Properties of Extruded or Worked Powder Alloys Compared
to Conventional Ingot Products

Alloy	Powder	Condition	Y.S. (0.2%) (psi)	U.T.S. (psi)	Elong. (%)	Remarks
18–8 0.03% S	Ingot	Annealed	35	80	65	
18–8 0.65% S	Coarse	As-extruded	80	120	55	Fine grained
	Steam	70% cold work	180	190	15	No interm. anneal
X–45	Prec. cast	As-cast	66	110	17	—
X–45	Coarse	HIP	84	163	28	235 ppm oxygen
	Steam	HIP	81	158	19	195 ppm "
Cu 0.74% Zr	Small ingot	Sol. treated +25% cold worked	46	50	5 (in 2")	
Cu 0.8% Zr*	Interm. powder	As-extruded	47	61	30	No cold work
Cu–Zr– Cr*	Interm. powder	Sol. treated + 75% cold work	73	76	14	85% IACS cond.

*Data from Reference [7].

show rates nearer $10^{2°}$C/sec; coarse steam-atomized powders (0.5-4 mm) will achieve rates of 10^{3}-$10^{4°}$C/sec. Cooling rates of 10^{3}-$10^{4°}$C/sec generally result in dendrite arm spacings of 5-10 microns, indicative of low segregation and a fine dispersion of excess phase particles.

A broad range of alloys have now been subjected to these higher cooling rates with excellent resultant properties. Table II lists comparative data for random alloys cooled at known indicated rates.

6. *Halide Decomposition.* The French ONERA* process utilizes a mixture of metal halide gases, for example, to produce an 18-8 Mo stainless steel (type 316), in powder form averaging perhaps 10 microns, and of close screen analysis. The alloy powders are formed in a porous packed bed of MgO; the alloy must contain chromium at a high enough content to permit subsequent leaching in acid to remove a residue of MgO. The purity of the material is quite high; however elements such as C, N, P, Si, and Mn are not present in alloying amounts. Various iron-, cobalt-, and nickel-base alloys containing at least 12 percent chromium can be produced. Figure 7 shows typical powder structures in this instance of an 18-8 Mo alloy. The pitted surface is apparently characteristic of the halide decomposition product.

*Office National d'Etudes et de Recherches Aérospatiales.

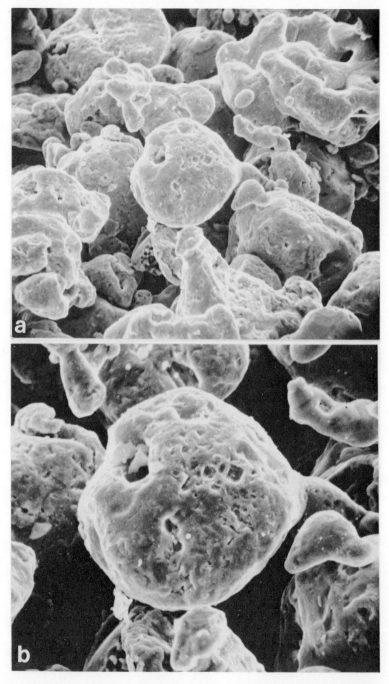

Figure 7. Scanning electron micrographs of French ONERA-type 18-8 Mo stainless steel powders made by halide decomposition technique. (a) is at 1000X, (b) is at 2200 X.

Summary

The six powder-producing methods described in this presentation are all capable of emerging as production methods. The powders vary from nearly micron-size particles to several millimeters in size; they offer different levels of purity, but basically are compatible with purity demands; they permit a broad range of alloy compositions; they appear to be of economical proportions. In the higher alloy compositions they appear attractive because of atmosphere control potential, rapid quenching (in some instances), and excellent yield of powder from ladle to screened product. Alloys have been produced and tested from these powders and, in the wrought form, offer improved properties compared to ingot product.

References

1. Bufferd, A.S. and Gummeson, P.U., "Application Outlook for Superalloy P/M Parts," *Metal Prog.*, 99 (1971), 68.
2. U.S. Patent No. 2,997,245, August 22, 1961, "Method and Device for Pulverizing and/ or Decomposing Solid Materials," E.O.F. Nilsson, S.I. Nilsson, and E.G. Hagelin to Kohlswa Jernverks Aktiebolag, Kolsva, Sweden.

   U.S. Patent No. 3,067,956, December 11, 1962, "Method and Device for Pulverizing and/or Decomposing Solid Materials," A. Tove to Kohlswa Jernverks Aktiebolag, Kolsva, Sweden.

   U.S. Patent No. 3,302,892, February 7, 1967, "Method and Device for Pulverizing Solid Material," E.O.F. Nilsson to Kohlswa Jernverks Aktiebolage, Kolsva, Sweden.

   Canadian Patent No. 703,416, February 9, 1965, "Method and Device for Pulverizing and/or Decomposing Solid Materials," A. Tov to Kohlswa Jernverks Aktiebolag, Kolsva, Sweden.
3. Daugherty, T.S., "Continuous Production of Aluminum Sheet from Finely Divided Particles," *Trans. AIME*, 233 (1965), 423.
4. Grant, N.J., "Structure and Property Control Through Rapid Quenching of Liquid Metals and Alloys," Proceedings of the International Conference on Metastable Metallic Alloys, *FIZIKA*, 2 suppl. 2 (1970), 16.
5. Duwez, P., "Bibliography on Alloys Quenched from the Liquid State," Proceedings of the International Conference on Metastable Metallic Alloys, *FIZIKA*, 2, suppl. 2 (1970), 1.
6. Lebo, M.R., "Structure Refinement and Property Control in Aluminum Alloys Through Quenching from the Melt," Unpublished Ph.D. dissertation, Massachusetts Institute of Technology, July 1971.
7. Sarin, V., Personal communication.

RECENT DEVELOPMENTS IN PRESSING AND SINTERING

MODERATOR: FREDERICK N. RHINES
University of Florida
Gainesville, Florida

6. Fundamentals of Sintering Theory

G. C. KUCZYNSKI

University of Notre Dame
Notre Dame, Indiana

ABSTRACT

The phenomenon of sintering, which manifests itself by coalescence of particles and shrinkage of pores in the powder compacts, is motivated by capillary forces. These forces actuate mass flows which tend to enlarge contact areas between adjacent particles and decrease the pore volume. Model experiments with the spherical particles, wires, and small tubes strongly indicate that the predominant flow is diffusional, such as that occurring in Nabarro-Herring creep. There is no direct experimental evidence that the creep by dislocation plays any important role in this process. Although simple models composed of particles of well-defined geometry can be fairly well understood, the actual mechanisms involved in sintering of powder compacts are very difficult to analyze. However, some modified models of high-density compacts can be, and were, discussed with some degree of success. The main problem seems to be the interdependence of pore shrinkage and grain growth. (The grain boundaries are the sinks for vacancies in diffusional flow.) Taking the well-known Zener's relation between pore or inclusion and grain diameter as an equation of grain growth, an equation linking porosity of a compact with time and temperature of sintering can be derived. It is in fair agreement with observations. The capillary forces are very weak; therefore, their importance decreases when other gradients such as concentration gradients are present. This has been demonstrated by model experiments on the polycomponent systems.

Introduction

The term "sintering" is commonly used to refer to the annealing treatment by which powders are consolidated into coherent and/or dense polycrystalline aggregates. The technological importance of this process hardly needs emphasizing. It is one of the most important and versatile methods of integration of materials in metallurgy and ceramics. Fast development of the art of powder technology in modern times necessitated better scientific understanding of the processes involved in sintering, the central problem of powder technology.

Even the simplest experiments performed on the powder compacts, such as measurement of volume changes during sintering, led to the conclusion that these changes must be a manifestation of some unbalanced forces acting between particles in the interior of the compacts. From the outset, chemists and engineers working in this field suspected that these forces are the same as those which cause spheroidization of water droplets and climb of liquids in capillary tubes—in short, surface or capillary forces which try to reduce total surface, and consequently the energy, of the system. Indeed, the mass of powder contains extra energy which can be released when the powder compact is converted into a solid body. The difference in rates of spheroidization of water droplets and that of densification of metal powder compact, is due to the difference in viscosities. Even close to melting point, solid metals have viscosities orders of magnitude higher than that of liquid water. The first section of this chapter will be devoted to a brief review of the nature of capillary forces and their importance in the process of sintering.

Capillary Forces

It is usually assumed that at equilibrium the pressures of two phases in contact should be equal. This is true if surface effects are not taken into consideration. However, if the surface of separation is not plane, then any displacement producing volume change will, in general, change its area and hence its energy. In other words, the existence of a curved surface of separation between two phases results in the appearance of additional force. As a result, the pressures in the two phases are not equal; the difference is called surface or interface pressure. This pressure can be found from a simple condition: that any change in surface energy produced by change of area dA, γdA, where γ is surface tension considered as isotropic, must be equal to the work of this pressure or stress σ, producing volume change dV. Thus,

$$\sigma dV = \gamma dA$$

or   (1)

$$\sigma = \gamma \frac{dA}{dV}$$

According to the well-known theorem in differential geometry,

$$\frac{dA}{dV} = \frac{1}{r_1} + \frac{1}{r_2}$$ (2)

where r_1 and r_2 are two principal radii of curvature of a surface at a given point. Equations (1) and (2) yield the well-known Laplace equation for the surface or interface pressure

$$\sigma = \gamma \left(\frac{1}{r_1} + \frac{1}{r_2} \right) \tag{3}$$

It is obvious that for a plane surface, $r_1 = r_2 = \infty$, and $\sigma = 0$. In the case of a sphere, $r_1 = r_2 = r$; therefore, $\sigma = 2\gamma/r$, and in the case of spherical cavity of radius of curvature $-r$, $\sigma = -2\gamma/r$ tensile stress—which tends to decrease the spherical pore volume. It should be noted that the interior of a compact contains numerous cavities of negative radii of curvature and that the capillary forces acting on the surface of these cavities motivate the phenomena observed during sintering.

Existence of the stress defined by equation (3) produces the change of chemical potentials in the phases separated by a curved interface. In a single component system, the change in chemical potential $\Delta\mu$ is

$$\Delta\mu = RT\ln(a/a_0) = \sigma V = \gamma V \left(\frac{1}{r_1} + \frac{1}{r_2} \right) \tag{4}$$

If the phase under consideration is the vapor phase, which can be treated as an ideal gas, then equation (4) can be reduced to

$$\frac{\Delta p}{p_0} = \frac{\gamma V}{RT} \left(\frac{1}{r_1} + \frac{1}{r_2} \right) \tag{5}$$

first derived by Lord Kelvin; p_0 is the vapor pressure over a flat surface and Δp the difference of pressures over curved and flat interfaces. Equation (5) indicates that over the convex surfaces, when total curvature is positive, as in the case of sphere, the vapor pressure is increased, $\Delta p > 0$. This explains why the rate of evaporation from the small globules of liquid is faster than from a flat surface. Of course, the opposite is true for a concave surface, and the vapor pressure in a cavity is less than that over the flat surface, $\Delta p < 0$, and the vapor tends to condense.

The effect of stress on the condensed system could be obtained by reasoning similar to that used in the case of the gas phase. However, we shall resort to one appealing to the physical picture first used by Nabarro [1]. Under the tensile stress, volume increase is expected, which will manifest itself in the increase of vacancy concentration. If the free energy to produce one mol of vacancies in the absence of stress is g_v^0, then when stress σ is acting, $g_v = g_v^0 + \sigma V$ where V is the molar volume of solid or liquid. Thus, the fraction of vacancies X_v is

$$X_v = \exp \left(-\frac{g_v^0 + \sigma V}{RT} \right)$$

or

$$X_v = X_v^0 \exp\left(\frac{-\sigma V}{RT}\right)$$

where X_v^0 is the fraction of vacancies under a flat surface.

But $\sigma V < RT$; therefore, with the help of equation (4), we obtain

$$\frac{\Delta X_v}{X_v^0} = \frac{-\gamma V}{RT}\left(\frac{1}{r_1} + \frac{1}{r_2}\right) \tag{6}$$

where $\Delta X_v = X_v - X_v^0$. This equation predicts increase of vacancy concentration under the concave surface and its decrease under the convex one.

Equation (3) and its consequences (4), (5), and (6) form a base of any sintering theory. The unbalanced force acting on the curved surface must actuate flow tending to decrease this force. The flow caused by capillary force produces the changes of volume and shape observed during sintering processes.

The capillary forces, as we have seen above, are strongly dependent on the surface geometry. The latter is very complicated in ordinary powder compacts, due to the unknown distribution of particle shapes and sizes; therefore, the general physical description and prediction of the changes taking place during sintering, utilizing equation (3), is hopeless—at least before the problems involving systems of simplified geometry are solved. Indeed, we owe all our understanding of sintering kinetics to the study of such simple models of well-defined geometry as spheres and cylinders, or combinations of spheres, cylinders, and plates. The review of the results obtained from the study of the model systems will be given in the next section.

Model Experiments of Sintering

The mass flows which can bring about the changes of volume and shape under the action of capillary forces are: viscous or plastic flow, volume diffusion [1,2,3], surface or grain-boundary diffusion, and evaporation and condensation. Let us consider the first stage of sintering, which can be best represented by the welding of two spheres together, as represented schematically in Figure 1, (a and b). The neck which forms between the spheres has two radii of curvature, internal x and external $-\rho$. From the geometry of the system, ρ is approximately either $x^2/2a$ or $x^2/4a$, depending on whether the spheres are tangent [Figure 1(a)] or interpenetrating [Figure 1(b)] during the process of sintering (a is the radius of the spherical particle). In the case of interpenetration, the distance between the particles decreases with time; material flowing from inside the system fills the space between two spheres. In the case of tangency, there is no appreciable shrinkage. Material filling the neck space is taken from the sphere's surface.

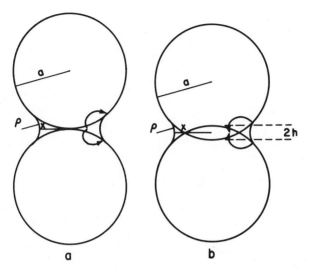

1. Schematic representation of two spheres or wires sintered together: (a) spheres tangent; (b) interpenetrating spheres.

Therefore, it follows that surface diffusion or evaporation and condensation cannot cause densification of compacts but, by building the bridges between adjacent particles, can increase their mechanical strength. Identification of the flow mechanism is done by mathematical analysis of the flow leading to the equation linking some easily measurable variables, such as neck radius x, time t, and temperature T. Then this equation is compared with the experiment. This has been done [2], and rather simple relations were obtained for initial sintering of sphere to sphere or plate and wire to wire or plate. The results are given in the following equation:

$$\left(\frac{x}{a}\right)^n = \frac{F(T)}{a^m} t \qquad (7)$$

where a is the radius of a sphere or wire and $F(T)$ a function of temperature only, which contains the proper coefficient characterizing given flow. Thus, when diffusional flow predominates, $F(T)$ contains the diffusion coefficient; in the case of evaporation and condensation mechanism, the vapor pressure p_0 of the solid; and when predominant flow is viscous, $F(T)$ contains viscosity. The actual predominating mechanism is characterized by exponents n and m in the following fashion:

when

$$
\begin{array}{lll}
n = 2 & m = 1 & \text{flow is Newtonian viscous} \\
n = 3 & m = 2 & \text{evaporation and condensation}
\end{array}
$$

$n = 5$ $m = 3$ volume diffusion
$n = 6$ $m = 4$ grain-boundary diffusion
$n = 7$ $m = 4$ surface diffusion

Numerous model experiments have been performed in which x was measured as a function of time. These measurements were done on cross sections through the wires or spheres sintered to large cylinders of the same materials, or to each other, as exemplified in the photographs reproduced in Figure 2. The results indicate that the predominant mechanism during the first stage of sintering of relatively large particles of metals and oxides at high temperatures is volume diffusion or Nabarro-Herring creep [2,4,5,6,7]. At lower temperatures and for small particles, surface-diffusion flow was often found to prevail [2,8,9,10]. Kinetics of welding of spheres of materials with high vapor pressure, such as NaCl, obeyed the low characteristic of the evaporation-condensation mechanism [5,11]. The beads of glass sinter by Newtonian viscous flow [12]. It should be noted that in the case of sintering of metals, glasses, and NaCl spheres, the diffusion coefficients, viscosities, and vapor pressures calculated from the neck growth rates were in very good agreement with the values of these coefficients determined by conventional methods. It should be added that in numerous model experiments performed with various crystalline materials, no evidence for plastic flow has been found. Some authors [13,14] regard equations (7) as only first approximations and attempted to arrive at more precise models by elaborate mathematical methods or computer calculations.

In some solids, like plastics, the flow is non-Newtonian. It can be described by the well-known Ostwald equation

$$\dot{\epsilon} = K\sigma^n \tag{8}$$

where $\dot{\epsilon}$ is shear strain rate and $n > 1$. The sintering equations for such solids takes the form [15]

$$\left(\frac{x^2}{a}\right)^n = \frac{n}{2}\left(\frac{8K}{n}\right)^n t \tag{9}$$

This equation was found to describe the sintering kinetics of poly (methyl methacrylate) [15] very well.

Equation (9) was used by E. Lenel [16], to rationalize sintering by dislocation creep. Using a Weertman [17] equation of the type of equation (8), with $n = 4.5$ to describe creep, one can attempt to compare the rates of neck growth, dx/dt, as predicted from diffusion and plastic flow theories. These two rates are: for plastic flow

$$\left(\frac{dx}{dt}\right)_{pf} = \frac{10^{-26}D}{a^3}\left(\frac{a}{x}\right)^9$$

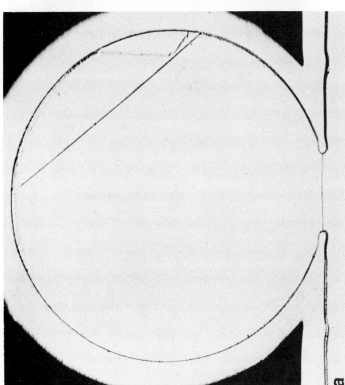

2. (a) Cross sections through copper wire sintered to copper cylinder at $T = 1050°C$ for time $t = 25$ hrs. 350X. (b) Spheres of Al_2O_3 sintered in hydrogen atmosphere at $T = 1825°C$ for time $t = 80$ min. 50X.

for diffusional flow

$$\left(\frac{dx}{dt}\right)_{vd} = \frac{1.7 \times 10^{-8} D}{a^2} \left(\frac{a}{x}\right)^4$$

where D is the self-diffusion coefficient of atoms of our solid. The plastic flow mechanism will be more efficient when $(dx/dt)_{pf} > (dx/dt)_{vd}$, which is fulfilled only when

$$\frac{x}{a} < 10^{-3} a^{-2/5}$$

Taking a typical value for the particle radius $a \approx 5 \times 10^{-3}$ cm, we come to the conclusion that the dislocation creep will prevail when $x/a < 10^{-2}$, or at a very early stage of sintering. However, this is not the whole story. In order to have creep by dislocations, we have to be sure that enough dislocations are in the neck volume to produce observed volume and/or shape change. Sintering is usually carried out at high temperatures and for relatively long time intervals, so one may expect that the dislocation concentration is at its lowest, as in a well-annealed metal—about $10^7/cm^2$. The neck area is approximately $2x\rho = \pi x^3/2a$. With $x/a < 10^{-2}$ and $a \approx 5 \times 10^{-3}$ cm, this area must be smaller than $4 \times 10^{-11} cm^2$. Therefore, the number of dislocations found in the volume enclosed by this area should be less than 4×10^{-4}! However, Lenel [17] points out that in the neck area or near the small pores, dislocations can be created at the grain boundary by the capillary stresses. Assuming this, we find that such a dislocation loop would require stress τ of the order of

$$\tau = \frac{Gb}{2x}$$

to be extended, where G is the modulus of rigidity, b the Burgers vector, and x the radius of the neck. The capillary stress acting on this loop is approximately $\sigma = \gamma/x$. Thus, in order that $\sigma > \tau$, $2\gamma > Gb$. It has been shown [18] quite generally that $2\gamma \leqslant Gb/4$; therefore, even if dislocations could be created in the neck area, capillary forces would be incapable of moving them. However, the most convincing experimental demonstration is provided by Brett and Seigle [19]. They sintered three twisted wires made of nickel with 2 volume percent of fine alumina particles dispersed in the metal. These particles served as inert markers. If plastic flow were a predominant mechanism of sintering, they should displace together with the metal and fill the neck cavity. Conversely, if the markers were not found in the cavities filled during sintering, the flow should be diffusional. The results of Brett and Seigle confirmed the second alternative. Indeed, one of their photographs, reproduced in Figure 3, clearly shows that the metal in the void between three nickel wires filled by sintering at 1400°C for 241 hours is

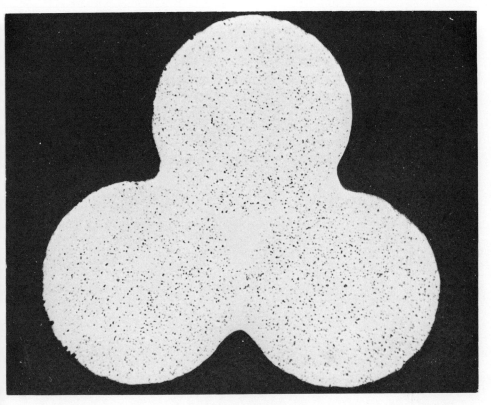

3. Three nickel wires sintered together at 1400°C for 241 hrs. The metal in the void is free of aluminum particles. 200X. (After Brett and Seigle [19].)

completely free of alumina particles. Due to the lack of resolution, this method cannot be used for the early stages of sintering.

Densification

The change of volume involved in densification of the powder compacts can be brought about only by the flow originating or ending in the interior of the particles. Such flows are: volume or grain-boundary diffusion, and viscous or plastic flows. Let us briefly consider volume diffusion. In order to fill a pore, vacancies have to be moved into its interior. Each atom entering a pore leaves a vacancy behind which moves in the opposite direction.

The process can be regarded as dissolution (vacancy by vacancy) of pore volume in the surrounding crystal. In order to shrink the volume of the system, all vacancies have to be eliminated either by diffusing them out of the system or destroying them in the internal sinks. The first alternative does not seem to be feasible because of the relatively large distances involved and the slowness of the diffusion process in the condensed phase. The second possibility requires the existence of effective vacancy sinks inside the crystals. Herring [3] was the first to suggest that grain boundaries are the effective sinks of vacancies, and Alexander and Baluffi [20] have verified this suggestion experimentally. They sintered a spool of copper wires and observed that tubular pores shrank uniformly as long as they were connected by a network of grain boundaries. When, by prolonged annealing, the grain boundaries were virtually eliminated, the shrinkage ceased.

This experiment explains old observations reported by powder metallurgists— that the powder compacts cannot be sintered to the full density by annealing alone. In order to eliminate this residual porosity, the compacts were often repressed and resintered. Repressing followed by reheating caused recrystallization, which restored the grain-boundary network connecting the pores. The experiment of Alexander and Baluffi dramatically emphasizes the important role grain boundaries play in densification.

The equation for tubular pore shrinkage by volume diffusion was derived by Kuczynski [21]

$$r_0^3 - r^3 = \frac{3\gamma VD}{RT} t \tag{10}$$

where r_o and r are radii of the pore at $t = 0$ and t, respectively; V is the molar volume of the material; and D its coefficient of self-diffusion. Model experiments of Ichinose [22] on the tubular pores formed between three twisted copper wires fully verified this equation.

Glasses, as has been mentioned above, sinter by Newtonian viscous flow. Work of Kuczynski and Zaplatynskyj [23] on shrinkage of glass capillary tubes verified the relation

$$r_o - r = \frac{\gamma}{2\eta} t \tag{11}$$

predicted for this process (η is the viscosity of glass).

Let us now turn to the application of the theory to sintering of actual powder compacts. Because of the complexity of the compact geometry, the theoretical treatment has to resort to simplified models.

Initial densification of a powder compact can be understood with the help of Figure 1(b). The linear shrinkage is simply proportional to the amount of interpenetration h. If we simplify a compact to an assembly of spheres of equal radius a, regularly distributed throughout the volume of the compact of linear dimensions L, then the linear shrinkage per unit length $\Delta L/L = h/a$. With our approximations, $h \cong \rho = x^2/4a$; thus, the shrinkage per unit length

$$\frac{\Delta L}{L} = \left(\frac{x}{2a}\right)^2$$

or, using equations (7), we obtain its dependence on time t and average particle size a

$$\frac{\Delta L}{L} = \frac{F(T)^{\frac{2}{n}} t^{\frac{2}{n}}}{4a^{\frac{2m}{n}}} \tag{12}$$

As only volume or grain-boundary diffusion flows are capable of changing the volume of the compact, and $F(T)_{vd}$ and $F_{bd}(T)$ are, respectively,

$$F(T)_{vd} = \frac{31\gamma\,VD}{RT}$$

$$F(T)_{bd} = \frac{96\gamma\,v\delta D_b}{RT}$$

where D and D_b are volume and grain-boundary diffusion coefficients, respectively, and δ the width of the grain boundary. The expressions for $\dfrac{\Delta L}{L}$ can be written as follows:

$$\left(\frac{\Delta L}{L}\right)_{vd} = \frac{\left(\dfrac{31\gamma VD}{RT}\right)^{2/5} t^{2/5}}{4a^{6/5}} \tag{13}$$

$$\left(\frac{\Delta L}{L}\right)_{bd} = \frac{\left(\dfrac{96\gamma V\delta Db}{RT}\right)^{1/3} t^{1/3}}{4a^{4/3}} \tag{14}$$

It is difficult to verify these equations, because particles in a compact are not perfect spheres; they are not of the same size and their stacking is imperfect. Bockstiegel [24] has shown that deviations from spherical shape can change the time-dependence in equations (13) and (14) quite drastically. However, some authors [5,25,26] claim that the time relations observed by them in metallic and oxide powder compacts, which more or less agree with equations (13) and/or (14), are meaningful. Perhaps!

Similarly, an expression for shrinkage per unit compact length for materials flowing viscously or plastically can be derived directly from equation (9). It is:

$$\frac{\Delta L}{L} = \frac{1}{4a}\left(\frac{n}{2}\right)^{1/n}\left(\frac{8\gamma K}{n}\right)t^{1/n} \tag{15}$$

No attempts have been made to verify these equations experimentally. [Equation (15) represents a family of equations with parameter n characterizing the rheological properties of the material.] According to the author of this review, a conclusion drawn from the agreement of shrinkage data and equation (15) that the predominant flow is plastic is unwarranted.

In derivation of equations (13), (14), and (15), the approximation for $\rho(= x^2/4a)$ was used. This approximation holds as long as $x/a < 0.3$; therefore, the validity of these equations is limited to $\Delta L/L$ less than 6 percent, or at very early stages of sintering only. The first attempt to derive an expression for the rate of shrinkage during the later stages of sintering was made by Coble [26]. He considered two stages, the intermediate and the final, which differ in details of geometry. In both cases, he assumed the matter transport to occur by diffusion.

The intermediate stage of sintering is associated with that group of structures which evolve during sintering in which the pore phase forms an essentially continuous cylindrical channel along the three-grain edges of the polyhedral grains. Therefore, the grains in his theory are replaced by some regular polyhedra of uniform volume. The choice of polyhedron is not essential, as long as its infinite set can completely fill the space. In such a model Coble can define porosity

$$P = \frac{V_{\text{pore}}}{V_{\text{pore}} + V_{\text{grain}}}$$

as a function of the edge length a of the chosen polyhedron. With these assumptions and general equations for diffusional fluxes, Coble was able to derive an equation for the rate of porosity change:

$$\frac{dP}{dt} = -\frac{10\gamma VD}{a^3 RT} \tag{16}$$

If we assume that a does not change with time, then by integration of equation (16) we obtain that porosity decreases linearly with time. Such a relation was never observed in actual compacts, because as the pores shrink, the grains grow. Consequently, a increases with time. Coble's measurements of grain growth in Al_2O_3 compacts indicated that $a^3 \cong Bt$, where B is a function of temperature only. Inserting this expression into Equation (16) and integrating, he obtained

$$P_o - P = \frac{10\gamma VD}{BRT} \ln \frac{t}{t_i} \tag{17}$$

where t_i is some initial time at which $P = P_o$. This equation compared favorably with his density measurements obtained on Al_2O_3 compacts. However, this agreement may be fortuitous. Other workers [17,28,29] working with metal or metal-oxide compacts obtained empirical relations of the form

$$\frac{1}{P^n} - \frac{1}{P_0^n} = \alpha t \tag{18}$$

where α is a function of temperature and exponent n usually varies between 2 and 3. The most serious objection to the Coble model is not that it does not agree with the experimental data but that it contains an internal inconsistency: a^3 in equation (16) cannot be simply replaced by Bt found from some experiments, because the rate of grain growth in powder compacts is intimately connected with the rate of the disappearance of porosity. The pores act as inclusions inhibiting grain growth. Zener [30] has derived a relation between average grain diameter a, in equilibrium with volume fraction of inclusions (in our case porosity P), and average inclusion radius r

$$a \cong \frac{r}{P} \tag{19}$$

This equation should be regarded as the equation describing growth of grains in powder compacts. On the basis of this assumption, the author of this review [31] was able to derive an equation similar to equation (18). It also follows from this theory that whenever porosity change with time is given by an expression of type (18), the grain growth is approximately described by the equation

$$a^n - a_0^n = r_0^n \alpha t \tag{20}$$

where a_0 and r_0 are the initial average grain and pore radii, respectively. Indeed, the results of work quoted above [27,28,29] and Coble's work on copper compacts seem to verify these relations.

Multicomponent Systems

Sintering of aggregates composed of particles of two or more components is part of the theory of solid-state reactions. A brief review of these processes will be given here, with the emphasis put on the modifications of the capillary forces introduced by the concentration gradients.

Let us consider sintering of two spheres of the same radius but composed of two different elements, A and B. As soon as a small interface forms between these two spheres, the interdiffusion starts. There are two fluxes. One flux is the vacancy flux due to capillary forces, discussed above; another, the interdiffusion flux due to the composition gradients.

As the diffusion coefficients of two interdiffusing species, D_A and D_B are generally not the same [33], there will be additional flux of vacancies due to interdiffusion. From the geometry of the system, it is obvious that the concentration gradients extend over a small distance, about 2ρ. Assuming that $D_A >$

D_B, the flux of vacancies due to the interdiffusion J_i is approximately

$$J_i = (D_A - D_B) \frac{\Delta C_0}{2\rho}$$

where ΔC_0 is the difference of concentrations across the neck which, in the case of components A and B being pure metals, should be close to unity. Thus, the undisturbed growth of the neck between two particles will be possible as long as the vacancy flux due to capillary forces, which is approximately

$$J_c = \frac{\gamma V \overline{D}}{R T \rho^2}$$

is greater than the interdiffusion vacancy flux, or when

$$J_c > J_i$$

$\overline{D}$ stands for some average value of D_A and D_B, say $\overline{D} = D_A + D_B/2$. This condition can be reduced to:

$$\rho < \left(\frac{D_A + D_B}{D_A - D_B} \right) \frac{\gamma V}{RT}$$

or   (21)

$$\frac{x}{a} < 2 \left[\frac{D_A + D_B}{D_A - D_B} \left(\frac{\gamma V}{aRT} \right) \right]^{1/2}$$

Taking again a typical value for $a \approx 5 \times 10^{-3}$ cm, and for γ about 1.5×10^3 dyne/cm, and considering that $(D_A + D_B)/(D_A - D_B)$ may change between 2 and 10 only, we obtain that sintering will continue up to the values for x/a ranging between 10^{-2} and 10^{-1}. Equation (21) emphasizes the weakness of the capillary forces.

Overwhelming flux of vacancies across the interface causes the change of neck geometry. Indeed, the neck volume becomes supersaturated with respect to vacancies. As in any supersaturated alloy, these vacancies have to precipitate out in the form of voids. The most natural nucleating site for vacancy precipitation is in the neck cavity. There cavities become critical when

$$\rho > \frac{2\gamma V}{RT} \frac{C_0}{\Delta C}$$

where C_0 and ΔC are equilibrium and excess vacancy concentrations, respectively. Precipitation of these vacancies enlarges the neck radius ρ, forming grooves of small curvature. As the diffusion fluxes tending to enlarge neck diameter are proportional to the curvature, the rate of sintering decreases. These effects were first observed by Kuczynski and Alexander [34]. One of the

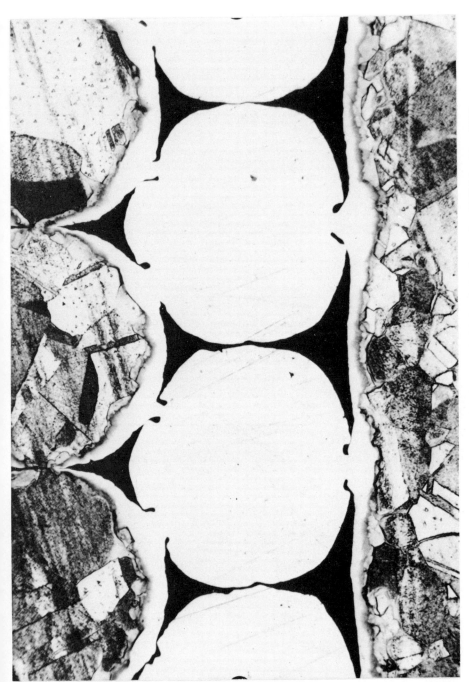

4. Gold wires (unetched) sintered to nickel wires and nickel cylinder (bottom) at 900°C for 1 hr. 200X.

photomicrographs showing grooving in the neck region is reproduced in Figure 4. A more thorough study of this problem on metallic systems has been done by Kuczynski and Stablein [35,36], and on oxide systems by Venkatu and Kuczynski [37].

References

1. Nabarro, F.R.N., "Deformation of Crystals by the Motion of Single Ions," *Report on a Conference on the Strength of Solids*, London: The Physical Society (1948), 75.
2. Kuczynski, G.C., "Self-Diffusion in Sintering of Metallic Particles," *Trans. AIME*, 185 (1949), 169.
3. Herring, C., "Diffusional Viscosity of a Polycrystalline Solid," *J. Appl. Phys.*, 21 (1950), 437.
4. Ichinose, H. and Kuczynski, G.C., "Role of Grain Boundaries in Sintering," *Acta Met.*, 10 (1962), 209.
5. Kingery, W.D. and Berg, M., "Study of the Initial Stages of Sintering Solids by Viscous Flow, Evaporation-Condensation, and Self-Diffusion," *J. Appl. Phys.*, 26 (1965), 1205.
6. Kuczynski, G.C., Abernethy, L. and Allan, J., "Sintering Mechanisms of Aluminum Oxide," *Kinetics of High Temperature Processes*, W.D. Kingery, ed., Cambridge, Mass.: Massachusetts Institute of Technology Press (1959), 163.
7. Coble, R.L., "Initial Sintering of Alumina and Hematite," *J. Am. Ceram. Soc.*, 41 (1958), 55.
8. Pranatis, A.L. and Seigle, L., "Sintering of Wire Compacts," in *Powder Metallurgy*, W. Leszynski, ed., New York: Interscience Publishers (1961), 53.
9. Fishmeister, H. and Zahn, R., *Ber. II Int. Pulver Met.*, Eisenach (1961), 93.
10. Matsumura, G., "Sintering of Iron Wires," *Acta Met.*, 19 (1971), 851.
11. Moser, J.B. and Whitmore, D.H., "Kinetics of Sintering Sodium Chloride in the Presence of an Inert Gas," *J. Appl. Phys.*, 31 (1960), 488.
12. Kuczynski, G.C., "Study of the Sintering of Glass," *J. Appl. Phys.*, 20 (1949), 1160.
13. Berrin, L. and Johnson, D.L., "Precision Diffusion Sintering Models for Initial Shrinkage and Neck Growth," *Sintering and Related Phenomena*, G.C. Kuczynski, N.A. Hooton, and C.F. Gibbon, eds., New York: Gordon and Breach, Science Publishers (1967), 369.
14. Nichols, F.A. and Mullins, W.W., "Morphological Changes of a Surface of Revolution Due to Capillary-Induced Surface Diffusion," *J. Appl. Phys.*, 36 (1965), 1826.
15. Kuczynski, G.C., Neuville, B. and Toner, H.P., "Study of Sintering of Poly (Methyl Methacrylate)," *J. Appl. Polymer Sci.*, 14 (1970), 2069.
16. Lenel, F.V., "The Early Stages of the Mechanism of Sintering," *Powder Metallurgy for High-Performance Applications*, J.J. Burke and V. Weiss, eds., Syracuse, N.Y.: Syracuse University Press (1972).
17. Weertman, J., "Theory of the Steady-State Creep Based on Dislocation Climb," *J. Appl. Phys.*, 26 (1955), 1213; and "Steady-State Creep Through Dislocation Climb," *J. Appl. Phys.*, 28 (1957), 362.
18. Kuczynski, G.C., "Some Relations Between the Modulus of Rigidity and the Surface and Sublimation Energies," *J. Appl. Phys.*, 24 (1953), 1250.
19. Brett, G.J. and Seigle, L. "The Role of Diffusion Versus Plastic Flow in the Sintering of Model Compacts," *Acta Met.*, 14 (1966), 575.

20. Alexander, B.H. and Baluffi, R.W., "The Mechanism of Sintering of Copper," *Acta Met.*, 5 (1957), 666.
21. Kuczynski, G.C., "The Mechanism of Densification During Sintering of Metallic Particles," *Acta Met.*, 4 (1956), 58.
22. Ichinose, D., "Sintering of Copper," Unpublished Ph.D. dissertation, University of Notre Dame, 1960.
23. Kuczynski, G.C. and Zaplatynskyj, "Sintering of Glass," *J. Am. Ceram. Soc.*, 39 (1956), 349.
24. Bockstiegel, G., "On the Rate of Sintering," *Trans. AIME*, 206 (1956), 580.
25. Johnson, D.L. and Cutler, I., "Diffusion Sintering: I—Initial Stage Sintering Models and Their Application to Shrinkage of Powder Compacts," *J. Am. Ceram. Soc.*, 46 (1963), 541; and "Diffusion Sintering: II—Initial Sintering Kinetics of Alumina," *J. Am. Ceram. Soc.*, 46 (1963), 545.
26. Coble, R.L., "Sintering Crystalline Solids: I—Intermediate and Final State Diffusion Models," *J. Appl. Phys.*, 32 (1961), 787; and "Sintering Crystalline Solids: II—Experimental Test of Diffusion Models in Powder Compacts," *J. Appl. Phys.*, *32* (1961), 793.
27. Bruch, C.A., "Sintering Kinetics of the High Density Alumina Process," *Am. Ceram. Soc. Bull.*, 41 (1962), 799.
28. Evans, P.B. and Ashall, D.W., "Grain Growth in Sintered Nickel Powder," *Int. J. Powder Met.*, 1 (1965), 32.
29. Bannister, M.J., "Interdependence of Pore Removal and Grain Growth During the Late Stages of Sintering Beryllium Oxide," *Sintering and Related Phenomena,* G.C. Kuczynski, N.A. Hooton, and C.F. Gibbon, eds., New York: Gordon and Breach, Science Publishers (1967), 581.
30. Zenner, C., in Smith, C.S., "Grains, Phases, and Interfaces: An Interpretation of Microstructure," *Trans. AIME*, 175 (1948), 47.
31. Kuczynski, G.C., to be published.
32. Coble, R.L. and Gupta, T.K., "Intermediate Stage Sintering," *Sintering and Related Phenomena*, G.C. Kuczynski, N.A. Hooton, and C.F. Gibbon, eds., New York: Gordon and Breach, Science Publishers (1967), 423.
33. Smigelskas, A.S. and Kirkendall, E.O., "Zinc Diffusion in Alpha Brass," *Trans. AIME*, 171 (1947), 130.
34. Kuczynski, G.C. and Alexander, B.H., "A Metallographic Study of Diffusion Interface," *J. Appl. Phys.*, 22 (1951), 344.
35. Kuczynski, G.C. and Stablein, P.F., "Sintering in Multicomponent Systems," *4th International Symposium on Reactivity of Solids*, Amsterdam: Elsevier Publishing Company (1960), 91.
36. Stablein, P.F. and Kuczynski, G.C., "Sintering of Multicomponent Metallic Systems," *Acta Met.*, 11 (1963), 1327.
37. Venkatu, D.A. and Kuczynski, G.C., "Sintering of Two Component Oxide Systems with Compound Formation," *Kinetics of Reactions in Ionic Systems*, T.J. Gray and V.D. Frichette, eds., New York: Plenum Press (1969), 316.

7. The Early Stages of the Mechanism of Sintering

F. V. LENEL

Rensselaer Polytechnic Institute
Troy, New York

ABSTRACT

The mechanism of material transport during the early stages of sintering is considered from both an experimental and a theoretical point of view. Experiments specially designed to distinguish between material transport by plastic flow, *i.e.*, dislocation motion, and by diffusional flow are presented. These include model experiments in which particles or wires containing as markers a finely dispersed second phase and experiments in which single-crystal wires of zinc are sintered together. The movement of the second phase in the marker experiments and the width of the neck as a function of their relative orientation in the zinc single-crystal wire experiments indicate that material transport by plastic flow predominates in the early stages of sintering.

Making certain assumptions, rate laws for the rate of neck growth and the rate of shrinkage during sintering by plastic flow are derived. The transition stress between sintering by plastic flow and by diffusional flow for a model system consisting of silver spheres is calculated as a function of sphere diameter, neck diameter–sphere diameter ratio, and temperature. The calculated value is compared with an experimentally determined value.

A model for the very earliest stage of contact formation between a wire and a flat is presented. In the calculation, the elastic energy to nucleate a dislocation half-loop of a given radius is balanced against the energy gained by annihilating the free surfaces which are bonded through dislocation nucleation. In this model, a sharp apex angle at the neck boundary is preserved. The rounded neck, which is formed during diffusion of atoms into this sharp tip, has a small radius of curvature and, therefore, a high stress exists at its surface due to surface-tension forces. The generation of dislocation at this neck and the movement into the interior of the neck are discussed.

Introduction

This chapter is concerned with the mechanisms of material transport in sintering. To discuss these mechanisms, a parallel is drawn between creep and sintering. The close relation of these processes was first observed by Pines [1]. Both involve changes in the shape of a solid under the influence of a stress at a temperature above one-half of its melting point. In creep, this shape change occurs under the influence of a constant externally applied stress; creep rates are generally studied on simple round or flat test specimens. When a compact pressed from powder is sintered, the changes in shape observed are neck growth between particles and shrinkage. Even when sintering is studied in a model experiment such as arrays of spheres or of wires, the shape changes are more complicated than in simple creep experiments. Sintering may occur under the influence of an external force as in hot pressing. In conventional sintering, however, the driving force is that of surface tension, and the stress due to this force is not constant, but decreases as sintering progresses.

There are still open questions regarding material transport in high-temperature creep, but there is general agreement upon the two principal mechanisms—one in the low-stress, the other in the high-stress region. In the low-stress region, material transport takes place by diffusional flow from boundaries under a compressive stress to boundaries under a tensile stress. Nabarro [2] and Herring [3] postulated a vacancy flux through the volume of the grains by lattice diffusion; Coble [4] postulated a flux through the grain boundaries by grain-boundary diffusion. In this type of viscous creep, the creep rate is proportional to the applied stress. It may occur under the influence of surface tension, as was shown in the experiments of Udin, Shaler, and Wulff [5]. In high-stress creep, on the other hand, the creep rate is generally proportional to a power of the stress, much higher than 1. In pure metals, this power is between 4 and 5. Weertman [6] has developed a model for this type of creep based upon the climb-controlled glide of dislocations. According to Weertman's model, and in agreement with experimental findings, the creep rate depends upon a hyperbolic sine function of the stress at very high stresses which, at somewhat lower stresses, reduces to a relation in which the creep rate is proportional to the 4.5th power of the stress.

In view of the large difference in the stress-dependence of the creep rate for diffusional creep and for creep by dislocation glide and climb, there must be a fairly sharp transition or transition range of stress below which diffusional transport mechanisms prevail and above which creep is controlled by dislocation motion. Since vacancies are generated and annihilated at grain boundaries in diffusional creep, the transition stress must depend upon the grain size of the material, decreasing with increasing grain size. This transition stress for creep in pure nickel with a grainsize of 0.033 cm has been estimated by McLean [7] to be approximately 1.3×10^7 dynes/cm^2.

The question of material transport in sintering was first studied by Frenkel [8], who postulated a viscous-flow mechanism. In his classical paper, "Self-Diffusion in Sintering of Metallic Particles," Kuczynski [9] added other possible mechanisms for material transport in sintering to the one postulated by Frenkel—

transport through the vapor phase, by volume self-diffusion and by surface self-diffusion. He also showed that it should be possible to distinguish between these mechanisms by measuring the rate of neck growth in a model experiment, in which a sphere is sintered to a plane. On the basis of his work with copper spheres sintered to a copper flat, Kuczynski [9] postulated a volume-vacancy diffusion mechanism, because his experiments showed that the fifth power of the neck radius is proportional to time. This mechanism of material transport by lattice diffusion of vacancies is analogous to the Nabarro-Herring mechanism of creep. The vacancy source is the surface at the periphery of the neck between sphere and flat, which has a small radius of curvature and is under high tensile stress due to surface-tension forces. Vacancy sinks are either the flat or nearly flat surfaces adjacent to the neck or the grain boundary between sphere and plane. Vacancy diffusion takes place under the influence of a gradient of stress which is also a gradient of chemical potential. Kuczynski [9] mentioned only in passing the possibility of material transport by plastic flow, *i.e.*, by dislocation motion. He postulated that in plastic flow, the square of the radius of the neck formed should be proportional to time in analogy to the similar relationship developed by Frenkel [8] for sintering by viscous flow. Kuczynski did not present any theoretical basis for this relationship.

The rate-law method has been widely used in order to study material transport in sintering. In addition to rate laws governing neck growth, laws were derived for the rate of shrinkage, *i.e.*, the rate at which the center of sintering particles approach each other. Yet another rate-law method is the "scaling factor" approach by Herring [10]. To the mechanisms considered by Kuczynski, transport by grain-boundary diffusion through the grain boundary in the neck was added. This mechanism corresponds to the one postulated by Coble [4] for creep. The rate-law methods of studying sintering mechanisms have distinct disadvantages. It is necessary to make assumptions regarding the geometry of the sintering array. Therefore, the rate laws derived are only approximations, as was shown by Nichols and Mullins [11] in the case of surface diffusion and by Easterling and Thölén [12] for volume diffusion. Furthermore, the exponent appearing in the rate law—*i.e.*, the exponent n in the relationship $x^n = t$, where x is the neck radius and t is time—and the analogous exponent in the rate law for shrinkage must be determined with considerable accuracy in order to distinguish clearly between mechanisms on the basis of rate laws. The problem is further complicated, as pointed out first by Rockland [13], when several mechanisms operate simultaneously and, as a result, the ratio of the contributions of the several processes varies continuously during the process.

In this presentation, the role of plastic flow (*i.e.*, dislocation motion) as a material transport mechanism in sintering is examined. It is postulated that early in the sintering process, when the radii of curvature of the free surfaces of the pores are very small and the stresses due to surface-tension forces are high, material transport by dislocation motion should predominate in analogy to the dislocation-motion mechanism of creep at high stresses. When the necks between particles grow and the pores spheroidize, the radii of curvature of the free surfaces increase, the stresses decrease, and material transport by diffusional flow should take over, again in analogy to the diffusional creep at low stresses. As in the case of creep, there should be a transition range between the mechanisms. As pointed out above, the most widely used way to distinguish between sintering mechanisms is by the rate-law methods. These methods are, however, not well suited to distinguishing between material transport by dislocation and by diffusional flow. For this reason, the first section of this presentation is devoted to a discussion of several experimental methods, specially designed for this purpose. On the other hand, since deriving relationships for the rate of shrinkage and the rate of neck growth is considered important in the understanding of a material-transport mechanism, such relationships for plastic flow are derived in the second section. As in the case of sintering by diffusional flow, assumptions must be made in order to derive the rate laws. The only previously suggested rate law for sintering by plastic flow is the one for neck growth by Kuczynski [9], which has been generally accepted, but is believed to be incorrect. Based on the newly derived rate law, the transition stress between transport by dislocation motion and by diffusion for silver is calculated. The results of the calculation are compared with an experimental determination of the transition stress.

There are important differences in the geometrical changes during material transport in creep and in sintering which have not yet been discussed. The possible effects of the size and the shape of the volumes within which plastic flow occurs during sintering upon the generation of dislocations and the development of dislocation substructure is discussed in the third section.

Experiments to Distinguish Between Plastic and Diffusional Flow

An experimental method which has been frequently used for providing evidence regarding the role of plastic deformation uses sintering with internal markers. The method consists of sintering together particles or wires containing as markers a finely dispersed second phase which is stable at the sintering temperature. The operative transport mechanism can then be determined in principle by studying the marker motion, which occurs during sintering. If material is transported by surface diffusion or volume diffusion with surface sources and sinks, no marker motion into the newly deposited neck-fillet regions will occur, since atom motion occurs by individual vacancy or atom jumps which could not move the markers. If transport occurs by diffusion, with the grain boundary between the particles as a vacancy sink, the center of the neck will contain markers, and marker-free zones will occur immediately under the neck surfaces where material is being deposited. The markers will accumulate on the grain boundary, since there is no place for them to go once they have arrived there. In the case where mass transport occurs by dislocation glide, the marker distribution in the deforming material must remain as it is with no concentration or depletion of marker density, and the neck regions will contain the same marker distribution as the areas from which the material is removed. Thus, by this method, it should be possible to distinguish plastic flow transport and diffusion of vacancies to the grain boundary from each other and from the remaining possible mechanisms.

Several such experiments [14-17] have been made, one of which [16] is illustrated by the photomicrographs, Figures 1-3. They show the necks between silver wires which contain a fine dispersion of 0.3-micron-diameter alumina particles. Figure 1 shows the array of wires before sintering; Figure 2 shows the array after 1 hour of sintering, when a neck is formed which contains marker particles across its entire width. Figure 3 shows the array after 240 hours of sintering. Regions free of markers are formed near the ends of the neck, but there is still a clearly distinguishable region containing markers in the center of

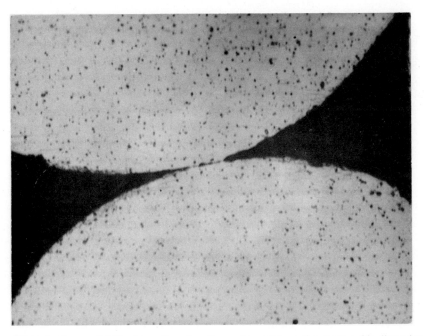

Figure 1. Array of 180-micron-diameter silver wires with an aluminum oxide dispersion before sintering. 630X.

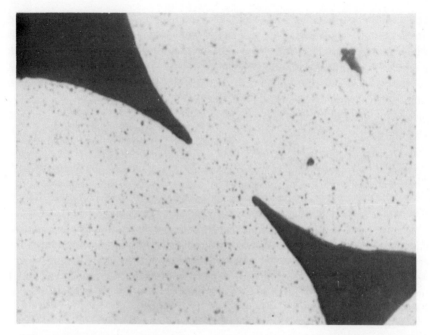

Figure 2. Array of 180-micron-diameter silver wires with an aluminum oxide dispersion after sintering 1 hr at 935°C. 580X.

the neck. When the marker-containing central zone is measured in specimens sintered for a series of times between 1 hour and 240 hours, it is found that its width stays approximately constant and is equal to the neck width after 1 hour of sintering. No marker pile-up is found at the grain boundaries in any of the specimens. These observations support a transport mechanism by dislocation motion during the first hour of sintering, rather than neck growth by diffusion of vacancies to the grain boundary between the wires. At longer sintering times, material transport by diffusional flow prevails.

Material transport by plastic flow early during sintering is most clearly demonstrated in an experiment in which zinc single-crystal wires, 250 microns in diameter, are sintered together [18]. The width of the neck between the wires is determined as a function of the relative orientation of the wires. The neck width varies, as shown in Figures 4 and 5, with the largest width more than twice as large as the smallest. The observed variation in neck width with relative orientation must be caused by the anisotropy of those properties of zinc which determine the rate of material transport during sintering. For sintering by diffusional flow, these are specific surface free energy and diffusivity; for sintering by evaporation and condensation, they are specific surface free energy and vapor pressure. Since approximate equations for the rate of neck growth exist for

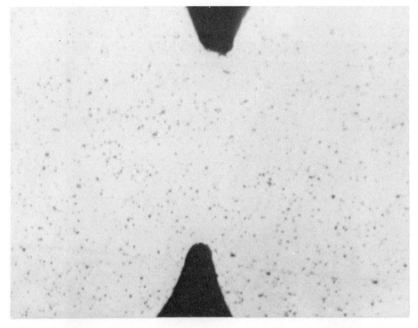

Figure 3. Array of 180-micron-diameter silver wires with an aluminum oxide dispersion after sintering 240 hrs at 935°C. 635X.

these mechanisms, the maximum ratios of neck width due to the anisotropies of specific surface free energies, diffusivities, and vapor pressures can be estimated. They are considerably smaller than the observed ratio of up to 2.7. On the other hand, the critical resolved shear stress on the basal slip system of zinc is five to ten times lower than for slip on any other system, even at temperatures near the melting point of zinc. The measured neck width of the single-crystal wire pairs can be directly related to the ease of basal slip. Those pairs which are oriented so that no basal slip is possible have small necks, while those which are favorably oriented for basal slip have large necks. Material transport by plastic flow must play an important role in the early stages of sintering of the zinc wires.

All of the sintering experiments discussed so far were concerned with isothermal sintering. In experiments by Morgan *et al.* [19–21], the shrinkage of compacts made of calcium fluoride and other ionically bonded materials was observed during heating rather than at constant temperature. The rates determined were much higher than would be predicted by diffusion models and were interpreted in terms of plastic-flow sintering. Also observed in these experiments was a temperature-dependent end-point in sintering shrinkage unexplainable by any diffusion model. In a transmission electron micrographic study of magnesium oxide compacts which had been given various sintering anneals,

Figure 4. Photomicrograph of zinc bicrystal oriented so as to preclude slip on the hexagonal base plane. 325X.

Morgan [22] showed large increases in the dislocation densities of the compacts associated with increased nonisothermal sintering rates, confirming his earlier conclusion that extensive plastic flow occurs during this type of sintering in many ionic systems.

Rate Laws for Sintering by Plastic Flow

The calculation of rate laws for models in which material transport takes place by diffusion are based on combining Kelvin's law—from which the gradient in vacancy concentration as a function of the radius of curvature is derived—with Fick's law for the rate of vacancy or material transport as a function of the

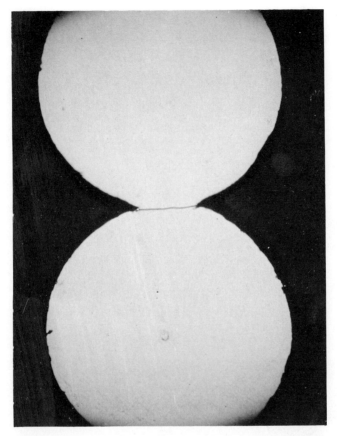

Figure 5. Photomicrograph of zinc bicrystal oriented so as to favor slip on the hexagonal base plane. 325X.

vacancy concentration gradient. In order to derive rate laws for material transport by plastic flow, the compressive stress acting in the neck between two spheres due to surface tension forces is calculated on the basis of a virtual work argument suggested by Lenel and Ansell [23]. This stress is related to the rate of deformation or strain rate, as the centers of the two spheres approach each other. The relationship between stress and strain rate is taken to be the one found in high-temperature creep of pure metals. The assumptions made in this derivation are pointed out as the steps in the derivation are presented.

In order to calculate the forces acting between the centers of the spheres, the model illustrated in Figure 6 is used. It is assumed that during sintering of the two spheres the volume of the fillet formed is equal to that of the interpenetrating lens. Model experiments in which silver spheres, 3–6 microns in diameter, were sintered together in the hot stage of an electron microscope [24] showed

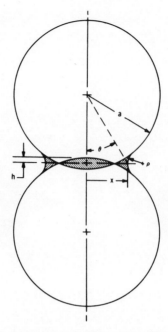

Figure 6. Two-sphere model with complete interpenetration.

such complete interpenetration only in a few cases. In the majority of the experiments the material in the fillet was formed at the expense of material from regions of the sphere surfaces adjacent to the neck, so that little or no approach of the sphere centers occurred during sintering.

To calculate the change in total surface free energy of the systems of two spheres, ΔG_s, the change in area of the system ΔA is multiplied by the specific surface free energy, γ, of silver. Values for the change in total surface area of the two-sphere model as a function of fractional interpenetration, h/a, are calculated [25] from the geometry of the model, assuming that the cross section of the neck surface with the radius ρ forms part of a circle, and that this surface is tangent to the surfaces of the two spherical particles. By differentiating the change in total surface free energy with respect to the center-to-center motion of the particles, the compressive force is obtained:

$$F = \gamma \, d\Delta A / dh \qquad (1)$$

This implies that the surface stress, *i.e.*, the force applied per unit length, is identical with the specific surface free energy. The specific surface energy of the grain boundary formed between the spheres is neglected.

In order to calculate the compressive stress acting between the spheres, the force F is divided by the cross-sectional area of the neck. This compressive stress is assumed to be uniform over the contact area between the two spheres. The results of the stress calculation can be expressed in the form of a dimensionless pressure coefficient P, which, when multiplied by the ratio γ/a (surface-tension-to-sphere diameter) results in the stress σ:

$$\sigma = P\,\gamma/a \qquad (2)$$

This dimensionless pressure coefficient P is found to be very nearly equal to the reciprocal of the fractional interpenetration h/a, another dimensionless parameter:

$$P = \frac{1}{h/a} \qquad (3)$$

This relationship applies to values of h/a up to 0.05 or 5-percent shrinkage. Since

$$\sigma\,a/\gamma = a/h \qquad (4)$$

$$\sigma = \gamma/h \qquad (5)$$

This provides a direct relationship between the stress and the amount $2h$ of interpenetration of the spheres or the thickness of the neck. Assuming that the strain in response to the compressive stress is essentially confined to a volume approximated by a disk with a radius equal to that of the interpenetrating lens, the velocity of the center-to-center motion, dh/dt, is set equal to $2h\dot{\epsilon}$, where $\dot{\epsilon}$ is the strain rate.

A rate law for shrinkage, i.e., for the rate of approach of the two spheres, can now be established by introducing Weertman's [6] relationship between stress and steady-state creep rate in pure metals:

$$\dot{\epsilon} = A\sigma^2 \sinh B\sigma^{2.5} \qquad (6)$$

Except at very high stresses, this relationship reduces to

$$\dot{\epsilon} = k\sigma^{4.5} \qquad (7)$$

The constants A and B, or the constant k, may be evaluated from measurements of steady-state creep of the metal in question in the appropriate stress range at constant temperature. Using equation (7), the rate of shrinkage may be written

$$dh/dt = 2h\,\dot{\epsilon} = 2hk\sigma^{4.5} = 2k\gamma^{4.5}/h^{3.5} \qquad (8)$$

This equation may be integrated:

$$h^{4.5} = 9k\gamma^{4.5}\,t \qquad (9)$$

In order to derive a rate law for neck growth, the half-thickness of the neck h

may be considered to be equal $x^2/4a$, where x is the neck radius. The rate law for neck growth becomes:

$$x^9/a^{4.5} = k't \qquad (10)$$

Introducing the stress–creep rate relationship for steady-state creep in the derivation of the rate law for shrinkage in the two-sphere model involves two assumptions. First, the relationship was derived for steady-state creep in pure metals, *i.e.*, for conditions in which the stress and strain rate are constant. In sintering they are not constant, but decrease continuously during neck growth. Second, the creep equation is predicated upon creep in bulk specimens, rather than upon the special geometry existing in sintering. This will be discussed in the third section.

The exponent 9 in the rate law derived for neck growth is much larger than the exponent 2 postulated by Kuczynski [9]. The exponent is also higher than those calculated for the diffusional mechanisms, which are 5 for volume diffusion and 7 for surface diffusion. This means that the initial sintering rate is faster and that it drops off more rapidly with neck growth. Rockland [13] has pointed out that in the case of simultaneous operation of two mechanisms, the exponential method of determining sintering mechanisms could result in an apparent exponent intermediate between the ones for the two mechanisms. Conceivably, the simultaneous operation of deformation by dislocation motion and by volume diffusion could be mistaken for surface diffusion. Since small particle size favors both surface diffusion and plastic flow, caution should be exercised in interpreting high neck-growth exponents as evidence of surface diffusion.

The rate of shrinkage when silver is sintered by plastic flow can be determined from equations such as equation (8). An actual calculation was performed using a modification of this equation, in which the more general strain rate–stress relationship of equation (6) is used. The parameters A and B were calculated from experimentally measured values of the stress- and temperature-dependence of the creep of pure silver.

This rate of shrinkage based on a plastic flow mechanism may now be compared with the rate of shrinkage of silver compacts by diffusional flow. For this purpose, the equations developed by Johnson [26] were used for a combined grain-boundary and volume-diffusion mechanism with suitable ranges of the parameters x/a (ratio of neck radius to sphere radius), the sphere radius a, and the temperature T. The results of this comparison are shown in Figure 7. The ratio x/a, at which the rate of plastic flow is equal to that of diffusional flow, is plotted as a function of the radius a for a number of temperatures. At these values of x/a, the transition from plastic flow to diffusional flow should occur.

Figure 8 is a plot of the stress at the transition value of x/a as a function of the particle radius. At the melting temperature, the transition stress varies from 10^9 dynes/cm^2 for 1-micron spheres to 10^7 dynes/cm^2 for 1-cm spheres. The

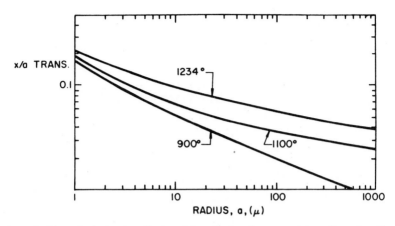

Figure 7. Plot of x/a (neck radius/particle radius) ratio vs. particle radius a for silver at the transition from plastic flow to diffusional flow at 900, 1100, and 1234°K.

higher transition stress for small particles is to be expected, since in sintering small particles the diffusion distances are smaller, the vacancy gradients are steeper, and the diffusional fluxes which transport material are greater than for large particles. As the temperature decreases, the transition stress increases. The reason is that the calculation of shrinkage rate by diffusional flow involves not only the activation energy for volume diffusion, which is the principal determinant for the temperature-dependence of shrinkage by plastic flow, but also the lower activation energy for grain-boundary diffusion.

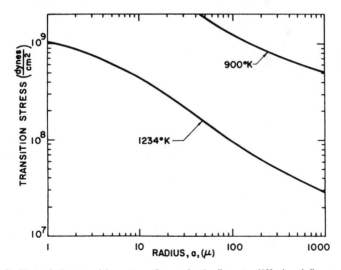

Figure 8. Plot of the transition stress from plastic flow to diffusional flow vs. particle radius in microns at 900 and 1234°K.

It is of interest to compare the transition stresses obtained by this calculation with one determined experimentally by Burr, Ansell, and Lenel [28]. In this experiment, the rates of neck growth were measured when a silver sphere is sintered to a flat plate of silver under the influence of external loads. Various external loads were used, which resulted in stresses large enough to outweigh the stress-contribution of the surface-tension forces. The neck-growth rate as a function of stress was determined by measuring these rates before and immediately after an incremental load was added. The neck-growth rates were taken to be proportional to the creep rates, $\dot{\epsilon}$, in creep-rate experiments. From the ratios of the neck-growth rates before and after the change in stress, the exponent n in the stress–creep rate equation was calculated from the formula:

$$n = \frac{\log \dot{\epsilon}_1/\dot{\epsilon}_2}{\log \sigma_1/\sigma_2} \qquad (11)$$

where $\dot{\epsilon}_1$ and $\dot{\epsilon}_2$ are the neck-growth rates before and after a load change, and σ_1 and σ_2 are the before and after values of the stress. The value of the exponent was found to be 1.8 for an average stress of 0.7×10^6 dynes/cm^2, and increased to about 4.5 at stresses of 5 to 10×10^6 dynes/cm^2. The transition stress measured in this experiment of approximately 4×10^6 dynes/cm^2 is about one order of magnitude lower than the one derived from the calculation for a 0.3-cm sphere at 800°C, which would be approximately 4×10^7, neglecting any difference due to the different geometry, sphere on sphere, rather than sphere on plate. One reason for this discrepancy may be the fact that the strain rates measured in the experiment immediately after an increase in stress were not steady-state strain rates but higher transient rates, while in calculating the stress exponent, steady-state conditions are assumed. The stress exponent measured may therefore be high and the transition stress low. On the other hand, the transition stresses derived from the calculation must be considered to be in reasonable agreement in view of the assumptions made in the derivation.

Dislocation Generation and Motion During Plastic-Flow Sintering

Weertman's mode [6] for steady-state creep is based on the assumption that an array of dislocation sources exists in the interior of the specimen subjected to a creep stress. Dislocations must be generated and, therefore, dislocation sources must also exist, when the strain necessary for sintering, *i.e.,* neck growth and shrinkage, is to take place by dislocation motion. The question to be considered in this last section is the location of these sources and the manner in which the dislocations generated from them move under the influence of surface-tension forces. The discussion is divided into two parts. In the first part, a model for the very earliest stage of contact formation is presented. The energy

necessary for forming a bond between a plate and a wire by generating dislocations is calculated. In this model, a sharp apex angle at the neck boundary is
preserved during contact formation. The second part deals with the situation
when a rounded neck has been formed by diffusion of atoms into the sharp tip
of the neck.

The calculation of the energy necessary for forming contacts was inspired by
observations on an analog system which consists of arrays of circular or flat rafts
of uniformly sized soap bubbles. As shown by Bragg and Nye [29], such rafts
deform plastically by dislocation generation and slip and by grain-boundary slid-
ing and migration in a manner quite similar to the one occurring in metal lattices.
The sintering of bubble rafts has been described previously [30], and illustrated
through a motion-picture film. Figure 9 shows two frames of this film which

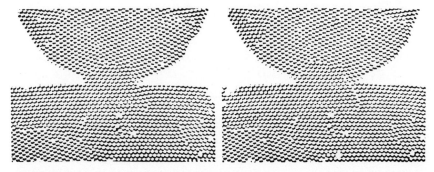

Figure 9. Sequence showing the nucleation of a dislocation at the right-hand apex of the
notch between a circular and a flat bubble raft.

demonstrate the sintering of a circle to a flat plate. A dislocation is nucleated in
the flat plate at the right side of the neck by punching in an extra half-plane of
bubbles which terminate at the neck boundary. The model used for calculating
[25] the energy necessary for bond formation in silver during sintering is illustrated in Figure 10. It consists of a plate and a wire, in which the contacting
surfaces are low-index planes so that they can be treated in terms of the terrace-
ledge-kink model. The elastic energy which must be expanded to nucleate a dislocation half-loop of a given radius by punching a segment of the first terrace
into either the plate or the wire is balanced against the energy gained by annihilating the free surfaces which are bonded through dislocation nucleation. The
radius of the dislocation loop depends upon the angle of the wedge-shaped notch
between plate and wire. For small angles, the radii of the dislocation loops are
quite large, e.g., at $900°$ for an angle of 0.10 radians; the radius is 10 microns.
The dislocations which are punched out can grow radially and can also expand
laterally. At small wedge-shaped notch angles, when the surface-energy term
exceeds the term for the elastic energy of the dislocation, this type of slip should

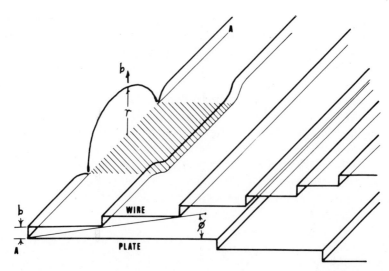

Figure 10. Nucleation of a dislocation half-loop at the tip of a wedge-shaped notch.

occur spontaneously. The calculation shows that for silver at $1200°K$, spontaneous growth should proceed for angles smaller than 0.162 radians, which corresponds to an x/a (neck radius/wire radius) ratio of 0.16. The calculation predicts bonding through generation of dislocations without thermal activation when two sufficiently clean, suitably oriented terrace-ledge surfaces are brought together at room temperature or below in the absence of an applied normal force. This process is, of course, entirely different from the thermally activated creep process of sintering postulated previously. It is interesting to note that it may very well be the process by which bond formation between contacting metal surfaces under conditions of extreme cleanliness occurs, as observed by Johnson and Keller [31].

The calculation of the energy of bonding in this process is based upon the assumption that a sharp apex is preserved at the neck boundary. The rate of nucleation of dislocations from the notch source will eventually decrease with increasing values of notch angle because of increased Taylor locking on the more closely spaced slip planes and because of the decreased driving force. When this happens, surface diffusion will round the sharp tip and eliminate the source.

In discussing the generation and motion of dislocations in the neck between two spherical particles having a geometry of the type shown in Figure 6, $i.e.$, a curved neck surface with a radius of curvature ρ, two related questions must be treated. One is the location of dislocation sources and the other the stress-distribution within the neck. Both questions have been discussed by Easterling and Thölén [32]. Their argument, briefly stated, is that the only possible location which would be under sufficiently high stress to generate dislocations would be the free surface. Away from the surface of the neck, the elastic stresses—as calculated by Easterling and Thölén—decrease rapidly to zero, so that even if dislocation sources existed in the interior of the neck, the stresses would be insufficient to generate dislocations. At the surface, the stresses would indeed greatly exceed the engineering yield stress of the material, but would not be high enough for dislocation generation. As an estimate of this stress, Easterling and Thölén use a value of $\mu/40$, where μ is the shear modulus of the material—in other words, a stress as high as the theoretical yield stress for perfect crystals. In order to develop a surface-tension stress of this magnitude, the radius of curvature, ρ, at the neck would have to be of the order of 10–20 Angstroms.

If the stress necessary to nucleate dislocations on the free surface is really $\mu/40$, nucleation would indeed be unlikely. This stress value is taken from a theoretical calculation of Hirth [33]. It may well be correct only within an order of magnitude. Hirth's discussion indicates that the details of the calculation strongly depend on the assumptions made regarding the structure of the surface, the effect of temperature, etc. A surface shear strength, an order of magnitude lower than the value assumed by Easterling and Thölén, would still be much larger than the bulk yield strength, but would make nucleation of dislocation at the surface of necks reasonable. Once dislocations are nucleated at the surface, they would glide under a net stress due to both the surface-tension stress and the stress field of dislocations subsequently nucleated. As a result, although the stresses at the surface would be decreased, the excess loading would be applied to areas in the particle interior, where the initial stress was lower. Easterling and Thölén's calculation of the initial elastic stress distribution would no longer apply, but the stresses in the interior of the neck would become sufficiently high to activate dislocation sources located there. If the total available load—as calculated by the virtual-work argument—is sufficient, the entire neck region will be maintained in a condition of steady-state creep as described by the Weertman model. Only when the average stress level decreases, as the neck grows and the transition stress level is reached, would transport by diffusional flow take over.

Conclusions

It is concluded that:

1. Experimental observations of the early stages of sintering between powder particles and model-system analogs strongly indicate that plastic deformation by dislocation motion is the predominant material-transport mode of sintering.

2. On the basis of the usual rate-law approach to sintering theory, a rate law for neck growth derived on the basis of climb-controlled dislocation motion yields the relationship $x^9 = k \cdot t$

3. Theoretical analysis shows that initial particle bonding by dislocation generation from the surface area at the apex of the sharp angle junction is energetically favorable.

4. The stresses developed during the early stages of sintering would be sufficient for both dislocation generation and subsequent glide.

Acknowledgment

The work reported in this contribution was supported by the Research Division of the U.S. Atomic Energy Commission at the NASA Interdisciplinary Materials Research Center of Rensselaer Polytechnic Institute. The contributions to this work of the author's colleague, Professor G. S. Ansell, and many of his students, in particular R. C. Morris and J. E. Sheehan, are gratefully acknowledged.

References

1. Pines, B. Ya., "Sintering (in the solid state)," *Sov. J. Tech. Phys.*, 16 (1946), 737.
2. Nabarro, F.R.N., "Deformation of Crystals by the Motion of Single Ions," *Report on a Conference on the Strength of Solids*, The Physical Society (1948), 75.
3. Herring, C., "Diffusional Viscosity of a Polycrystalline Solid," *J. Appl. Phys.*, 21 (1950), 437.
4. Coble, R.L., "Initial Sintering of Alumina and Hematite," *J. Am. Ceram. Soc.*, 41 (1958), 55.
5. Udin, H., Shaler, A.J. and Wulff, J., *Trans. AIME*, 185 (1949), 186.
6. Weertman, J., "Theory of Steady-State Creep Base on Dislocation Climb," *J. Appl. Phys.*, 26 (1955), 1213; and "Steady-State Creep Through Dislocation Climb," *J. Appl. Phys.*, 28 (1957), 362.
7. McLean, D., "Point Defects and the Mechanical Properties of Metals and Alloys at High Temperature," *Vacancies and Other Point Defects in Metals and Alloys*, London: The Institute of Metals (1958), 187.
8. Frenkel, Ya. I., "Viscous Flow of Crystalline Bodies Under the Action of Surface Tension," *J. Phys.*, USSR, 9 (1945), 385.
9. Kuczynski, G.C., "Self-Diffusion in Sintering of Metallic Particles," *Trans. AIME*, 185 (1949), 169.
10. Herring, C., "Effect of Change of Scale on Sintering Phenomena," *J. Appl. Phys.*, 21 (1950), 301.
11. Nichols, F.A. and Mullins, W.W., "Morphological Changes of a Surface of Revolution Due to Capillary-Induced Surface Diffusion," *J. Appl. Phys.*, 36 (1963), 1826.
12. Easterling, K.E. and Thölén, A.R., "Computer-Simulated Models of the Sintering of Metal Powders," *Z. Metallkunde*, 61 (1970), 928.
13. Rockland, J.G.R., "The Determination of the Mechanism of Sintering," *Acta Met.*, 15 (1967), 277.
14. Brett, J. and Seigle, L.L., "The Role of Diffusion Versus Plastic Flow in the Sintering of Model Compacts," *Acta Met.*, 14 (1966), 575.

15. Hingorany, A.R., Lenel, F.V. and Ansell, G.S., "The Role of Plastic Flow by Dislocation Motion in the Sintering of Calcium Fluoride," *Kinetics of Reactions in Ionic Solids,* T.J. Gray and V.D. Frichette, eds., New York: Plenum Press (1969), 375.

16. Lenel, F.V., Ansell, G.S. and Barron, D.D., "Marker Transfer During the Early Stages of Sintering," *Met. Trans.,* 1 (1970), 1772.

17. Lenel, F.V., Ansell, G.S. and Morris, R.C., "Sintering of Loose Spherical Copper Powder Aggregates Using Silica as Markers," *Met. Trans.,* 1 (1970), 2351.

18. Nunes, J.J., Lenel, F.V. and Ansell, G.S., "The Influence of Crystalline Anisotropy on Neck Growth During the Sintering of Zinc," *Acta Met.,* 19 (1971), 107.

19. Yust, C.S. and Morgan, C.S., "Material Transport During Sintering of Materials with the Fluorite Structure," *J. Nuc. Mater.,* 10 (1963), 182.

20. Morgan, C.S., McHargue, C.J. and Yust, C.S., "Material Transport in Sintering," *Proc. Brit. Ceram. Soc.,* Fabrication Science, No. 3 (1965), 177.

21. Morgan, C. S., "Densification Kinetics During Nonisothermal Sintering of Oxides," *Kinetics of Reactions in Ionic Solids,* T. J. Gray and V. D. Frichette, eds., New York: Plenum Press (1969), 349.

22. Morgan, C.S., "Mechanistic Interpretation of Non-Steady State Sintering," International Powder Metallurgy Conference, New York, July 1970, in *Modern Developments in Powder Metallurgy,* Vol. 4, *Processes,* H.H. Hausner, ed., New York: Plenum Press (1971), 231.

23. Lenel, F.V. and Ansell, G.S., "Creep Mechanisms and Their Role in the Sintering of Metal Powders," International Powder Metallurgy Conference, New York, July 1965, in *Modern Developments in Powder Metallurgy,* Vol. 1, *Fundamentals and Methods,* H.H. Hausner, ed., New York: Plenum Press (1966), 281.

24. Gessinger, G.H., Lenel, F.V. and Ansell, G.S., "Continuous Observation of the Sintering of Silver Particles in the Electron Microscope," *Trans. ASM,* 61 (1968), 598.

25. Lenel, F.V., Ansell, G.S. and Morris, R.C., "Theoretical Considerations and Experimental Evidence for Material Transport by Plastic Flow During Sintering," International Powder Metallurgy Conference, New York, July 1970 in *Modern Developments in Powder Metallurgy,* Vol. 4, *Processes,* H.H. Hausner, ed., New York: Plenum Press (1971), 199.

26. Johnson, D.L., "Sintering Kinetics for Combined Volume, Grain Boundary, and Surface Diffusion," *Phys. Sint.,* 1 (1969), B1.

27. Morris, R.C., "Dislocation Glide Deformation in the Sintering of Crystalline Solids," Unpublished Ph.D. dissertation, Rensselaer Polytechnic Institute, June 1972.

28. Burr, M.F., Lenel, F.V. and Ansell, G.S., "Influence of Pressure Upon the Sintering of Silver," *Trans. AIME,* 239 (1967), 557.

29. Bragg, W.L. and Nye, J.F., "A Dynamic Model of a Crystal Structure," *Proc. Roy. Soc.,* 190A (1947), 474.

30. Lenel, F.V., Ansell, G.S. and Morris, R.C., "A Bubble Raft Model to Study Sintering by Plastic Flow," *Advanced Experimental Techniques in Powder Metallurgy,* J.S. Hirschhorn and K.H. Roll, eds., New York: Plenum Press (1970), 61.

31. Johnson, K.I. and Keller, D.V. Jr., "Effect of Contamination on the Adhesion of Metallic Couples in Ultra-High Vacuum," *J. Appl. Phys.,* 38 (1967), 1896.

32. Easterling, K.E. and Thölén, A.R., "A Study of Sintering Using Hot-Stage Electron Microscopy," *Metal Sci. J.,* 4 (1970), 130.

33. Hirth, J.P., "The Influence of Surface Structure on Dislocation Nucleation," Conference on the Relation Between the Structure and Mechanical Properties of Metals, *National Physical Laboratory Symposium,* no. 15, Vol. 1, London: H.M. Stationary Office (1963), 218.

8. Recent Developments in the Analysis of Sintering Kinetics

D. L. JOHNSON
Northwestern University
Evanston, Illinois

ABSTRACT

There has been for many years a desire to determine the mechanism of sintering of metallic and ceramic materials and the effects of processing variables upon these mechanisms and upon the final product. In almost all studies of sintering mechanisms the data were obtained and interpreted in accordance with models which had so many simplifying assumptions in their derivation that great uncertainties exist as to the interpretation of the results. It has almost never been possible to predict, with any degree of certainty, the sintering behavior of any material. Sintering practice was and is a matter of cut and try. This chapter describes experimental procedures and methods of interpretation of the data which permit reliable conclusions concerning the mechanism of initial- and intermediate-stage sintering, even though several mechanisms may be occurring simultaneously. Perhaps the most useful of these methods is the study of microstructural evolution and sintering kinetics during the intermediate and final stages of sintering, inasmuch as these not only provide for determining the mechanisms but also allow for correlations between processing variables such as sintering atmosphere and additives with the microstructural evolution and, therefore, with the final properties. Although there are not sufficient data to do so at this time, it is likely that this approach will permit prediction of sintering behavior as we understand more about the relationships among the microstructural and mass-transport parameters.

Introduction

After many years of struggling with the phenomena involved in sintering, the simple models of Kuczynski [1,2] were a welcome development. It seemed that one could, at last, attach some physical significance to the phenomena that occur during sintering. The predominating mechanism for sintering was identified

139

by measuring the neck growth between particles as a function of time, and ascertaining the best value of n in the following equation:

$$X^n = K_1 t \qquad (1)$$

where X is the neck radius divided by the particle radius, K_1 is a constant containing the appropriate mass-transport parameters, and t is time. Kingery and Berg [3] extended the derivation to include shrinkage of a pair of spheres with the following result:

$$Y = (K_2 t)^m \qquad (2)$$

where Y is the fractional shrinkage, K_2 is a constant similar to K_1, and m is a constant. The values of m and n are presumed to identify the sintering mechanism. Several people have derived equations similar to these based upon similar or different boundary conditions [4], but most of the equations reduce to one of these with different values of the exponents and/or constants.

These equations have been used by many investigators to interpret sintering results. Unfortunately, most of these workers have failed to recognize that the simplifying assumptions which are required to derive these equations are very severe. Particularly troublesome is the assumption that a single mechanism is producing sintering. If two or more mechanisms are operating, the observed values of n or m will be altered but may still fit one of the values predicted by one of the models, leading to an erroneous conclusion. In addition, equation (2) has been employed frequently to analyze sintering of compacts of nonspherical particles, which is beyond its intended applicability. Other phenomena in the sintering of powder compacts, such as particle rearrangement and formation of new points of contact during shrinkage, make the sintering behavior of compacts considerably more complicated than equation (2) implies. Both equations (1) and (2) suffer from the necessity of knowing precisely the initial time, and also the initial length in the case of equation (2). Since the sintering experiment begins at room temperature, and a finite time is required to heat to the isothermal sintering temperature, this can lead to problems in accurately determining n or m, even if a single mechanism far overshadows all others. In addition, any particle rearrangement during heating of a compact will give an unexpectedly high shrinkage. Some workers have applied shrinkage and time corrections to their data to obtain agreement with the models [5-7]. However, at this time it appears that the variabilities in compacts and the uncertainties in the very derivations of the models severely undermine the confidence that one can place in either equation for analysis of sintering data.

Geometrically simple model systems have been extended into the intermediate stage by Coble and others [8-11]. Here again, however, a single mechanism was assumed and, more severely, a specified geometrical grain shape was assumed.

Other investigators have recognized the difficulty in clearly defining sintering mechanisms using the models and have sought alternative approaches. Tikkanen and Makipirtti [12] introduced the following empirical sintering equation which fitted much of their data:

$$\frac{V_0 - V}{V - V_{th}} = (K_3 t)^p \tag{3}$$

where V_0 is the initial volume of the compact, V is the volume at time t, V_{th} is the volume at theoretical density, K_3 is the rate constant, and p is a constant. Recognizing that the fractional linear shrinkage, Y, is approximately one-third the fractional volume shrinkage, this equation can be written in the following form:

$$Y = (Y_{th} - Y)(K_3 t)^p \tag{4}$$

where Y_{th} is the shrinkage at theoretical density. Thus for $Y \ll Y_{th}$, this equation reduces to the form of equation (2), although Tikkanen and Makipirtti attached no significance to the value of p. They compared the activation energy of K_3 with those for volume and grain-boundary diffusion to indicate densification mechanisms.

Still others have employed nonisothermal techniques in an attempt to understand sintering mechanisms [13–16]. Unfortunately, the apparent activation energy will be altered by concurrent mechanisms, leading to possible confusion [17]. One additional problem has been reported by Morgan [18,19], who observed that the apparent activation energy for densification of ThO_2 during a sudden increase in temperature is substantially greater than that for a sudden decrease.

An additional approach has been to follow the topology development of sintering bodies, rather than to be concerned about mechanisms [20–22]. The results are, of course, descriptive in nature.

As a consequence of the many models and different approaches, widely divergent views exist about the sintering of most materials that have been studied. Correlation of sintering phenomena with more fundamental parameters, such as diffusion coefficients, has been frequently less than satisfactory, although there have been exceptions. None of the models have been particularly useful in designing sintering practice and the accomplishment of industrial sintering goals. There are too many interrelated concurrent phenomena to permit simple interpretations.

As an example of the confused state of affairs, consider the case of silver. Although sintering studies on this material date back many years, the mechanisms of sintering are still subject to controversy. Kuczynski [1,2], using his simplified models for sintering of spheres and wires to a plane, concluded that silver sinters by volume diffusion. Johnson and Clarke [23] concluded from their studies of compacts of spherical particles that sintering occurs by a combination of grain-boundary and volume diffusion. They were forced to assume that surface diffusion was negligible. Nichols [24] reinterpreted Kuczynski's data on the sintering of wires and decided that all of the neck growth between the wires was produced by surface diffusion. Salkind *et al.* [25] investigated the sintering of loose silver powder near 300°C under an external load. By measuring the densification rate as the load was changed abruptly, they found that the densification rate was a power function of stress and inferred that plastic flow was responsible for densification. Burr *et al.* [26] studied the sintering of a sphere to a plate at 800°C under an external load and also stated that plastic flow was responsible for the neck growth, provided the stress was greater than 5×10^6 dyne/cm². Gessinger *et al.* [27] measured neck growth and center-to-center approach of micron-sized silver spheres on the hot stage of an electron microscope. They reported a predominance of surface diffusion at lower temperatures with an increase in shrinkage-producing mechanisms at higher temperatures. Because of contamination in the microscope, they were not able to interpret their data quantitatively. Gessinger [28] and Hirschhorn and Berglund [29] investigated the effect of oxygen in the atmosphere on sintering of silver. The former concluded that oxygen must enhance grain-boundary diffusion, while the latter stated that the enhanced densification was due to enhanced surface diffusion in the presence of oxygen. Herman [13] used his data on nonisothermal sintering of silver to conclude that a modified volume-diffusion mechanism was responsible. Finally, Seidel and Johnson [30] measured both the shrinkage and neck growth of compacts of spherical particles and observed that surface diffusion was indeed negligible, as Johnson and Clarke [23] have assumed, and that a combination of volume and grain-boundary diffusion produced densification.

The confusion that exists in the case of silver is by no means unique to this material. It is apparent that sintering must be studied in more detail in order to eliminate this kind of disparity of opinion. In particular, it is important that any sintering model employ as few assumptions as possible about particle geometry and mechanisms of mass transport. In addition, the experimental methods must include the kinds of measurements that will eliminate as many geometric assumptions as possible. Recent advancements in techniques do make it possible to determine sintering mechanisms with a minimum of uncertainty.

Nonisothermal Techniques

If the mechanism of mass transport in a compact of uniform spheres is either grain-boundary or volume diffusion, one of the following equations approximately describes the first 4.5 percent of linear shrinkage [31]:

$$TY^2\dot{Y} \cong \frac{0.78\,\gamma\Omega bD_b}{ka^4} \qquad (5)$$

$$TY\dot{Y} \cong \frac{2.4\,\gamma\Omega D_v}{ka^2} \qquad (6)$$

where $\dot{Y}$ is the fractional linear shrinkage rate, T is the absolute temperature, γ is the surface tension, Ω is the atomic volume, b is the width of the region of enhanced diffusion at the grain boundary, D_b is the boundary-diffusion coefficient, k is Boltzmann's constant, a is the particle radius, and D_v is the volume-diffusion coefficient. If grain-boundary diffusion is operating exclusively, for instance, a plot of the logarithm of the left-hand side of equation (5) versus $1/T$ would produce a single straight line as the temperature was increased and decreased in a cyclic fashion. If, on the other hand, surface diffusion is occurring along with grain-boundary diffusion, the initial heating period will show an abnormally steep line on this plot [17]. This is because surface diffusion retards densification, and surface diffusion is more important at lower temperatures and at lower shrinkages. At higher temperatures and shrinkages it becomes relatively less important and its influence on $\dot{Y}$ is reduced. Now, during a cool-down cycle the geometry changes only slowly, and the curve parallels the curve predicted for pure grain-boundary diffusion. Subsequent reheating follows the cooling curve until the temperature begins to approach that temperature at which the first cooling cycle began, at which time there are positive deviations. Figure 1 shows a curve which was synthesized by a computer for these conditions for silver. Experimental curves very similar to this have been determined for this material [31]. On the other hand, if volume diffusion plus surface diffusion was occurring, then a plot of the left-hand side of equation (6) versus $1/T$ would show a similar behavior, as shown in Figure 1.

This cyclic nonisothermal method is also useful in studying transient phenomena that occur in the very initial stages of sintering. For instance, if residual stresses are present in the particles, densification may be enhanced during the release of these stresses, or during recovery and recrystallization of the individual powder particles. Furthermore, plastic flow by dislocation motion would also result in transient enhanced densification. Any of these would give experimental curves on the plots of equations (5) and (6) which lie above those predicted for diffusion. Furthermore, during subsequent reheating the curves would lie below the initial heating curve. By comparing annealed powders, one could investigate the origin of this enhanced densification.

This approach is perhaps limited to the use of spherical particles, but the mechanism of densification and also the relative importance of surface diffusion can be determined with a minimum number of measurements. This experiment

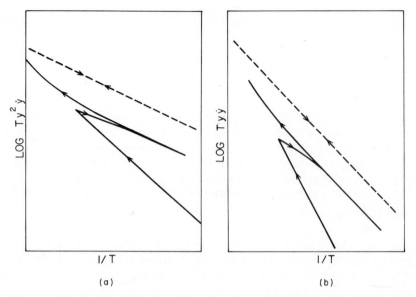

Figure 1. Computer synthesis of sintering data for cyclic heating and cooling: (a) grain-boundary and surface diffusion, plotted according to equation (5); (b) volume and surface diffusion plotted according to equation (8). Broken lines indicate only (a) grain-boundary diffusion and (b) volume diffusion.

is best conducted with a recording dilatometer, but corrections must be made for thermal expansion of the sample and the dilatometer.

Initial-Stage Isothermal Sintering

If two spheres of the same diameter, or a compact of spheres of the same diameter in which the number of contacts per sphere is independent of time, are sintering, the following equation relates the linear shrinkage rate to the instantaneous geometry [32];

$$\frac{X^3 R}{X + R} \dot{Y} = \frac{2\gamma\Omega D_v}{\pi k T a^3} \cdot \frac{A}{X} + \frac{4\gamma\Omega b D_b}{k T a^4} \tag{7}$$

where R is the minimum radius of curvature of the neck surface divided by the sphere radius and A is the surface area of the neck divided by a^2. This equation applies if X, R, and A are known, provided all densification is caused by volume and grain-boundary diffusion. Experimentally, one would measure a series of samples sintered at various times at a fixed temperature, and for each sample determine $\dot{Y}$, Y, and X. Knowing Y and X, R and A can be computed from the following equations:

$$R = \frac{Y^2 - 2Y + X^2}{2(1 - X)} \tag{8}$$

$$A = 4\pi R \left[\theta (X + R) \cos \theta - R \sin \theta \right] \tag{9}$$

where

$$\theta = \sin^{-1} \frac{1 - Y}{X + R} \tag{10}$$

By plotting the left-hand side of equation (7) versus A/X, the volume- and grain-boundary diffusion coefficients can be calculated from the slope and the intercept of the resulting curves, respectively. The ratio of grain-boundary to volume-diffusion flux is just the ratio of the second to the first term on the right-hand side. This equation applies even though surface diffusion and vapor transport are complicating the kinetics, inasmuch as the geometric parameters are measured explicitly or calculated from measured quantities. The surface-diffusion coefficient can be calculated by computer synthesis of the sintering curve. The results to date have given diffusion coefficients which are in very close agreement with published values [32].

If plastic flow is contributing to densification, equation (7) will not apply. However, it has been recognized that plastic flow should be transient in nature, since the stress in the neck diminishes as shrinkage proceeds [26]. When the stress falls below a certain value, densification will be by diffusion processes only. Beyond this point equation (7) will apply, whereas before it the actual densification rate, and therefore the left-hand side of equation (7), will be greater than that predicted on the basis of extrapolation in the diffusion-controlled range. Results indicative of this behavior have been obtained with lithium fluoride [33,34]. Furthermore, the lithium fluoride results, including the initial enhancement, have been substantiated by nonisothermal heating and cooling experiments [34].

Intermediate-Stage Isothermal Sintering

The geometric and mass-transport assumptions of the intermediate-stage sintering model referred to above have been overcome in a recent model which makes it possible to determine the volume and grain-boundary contributions to sintering, irrespective of complications due to surface diffusion, vapor transport, grain growth, pore growth, etc. [35]. The model equation is similar to equation (7), and reduces to equation (7) when the appropriate geometric parameters are substituted. It is as follows:

$$\frac{\overline{G}}{\overline{H}L_V} \frac{dL}{Ldt} = \frac{10\gamma\Omega D_v}{kT} \frac{S_V}{L_V} + \frac{10\gamma\Omega b D_b}{kT} \tag{11}$$

where $\overline{G}$ is the mean grain size, $\overline{H}$ is the mean surface curvature of the pore surfaces, L_V is the length per unit volume of grain boundary–pore intersections, S_V is the surface area of pores per unit volume of sample, and L is the sample length. Each of these parameters can be measured by straightforward quantitative microscopic techniques. The following relationships exist between these parameters and measured quantities:

$$\overline{G} = 1.5\,\overline{\ell} \qquad\qquad (12)$$

$$S_V = 2N_L \qquad\qquad (13)$$

$$L_V = 2N_A \qquad\qquad (14)$$

$$\overline{H} = \frac{\pi T_{A\,\mathrm{net}}}{N_L} \qquad\qquad (15)$$

where $\overline{\ell}$ is the mean grain intercept length for a test line on the sample polished surface, N_L is the number of intersections per unit length of test line with pore surfaces, N_A is the number of grain boundary–pore intersections per unit area on the plane of polish, and $T_{A\,\mathrm{net}}$ is the number of times a test line is tangent to convex portions of pore surfaces, minus the number of times it is tangent to concave segments of pore surfaces (per unit area) as the test line is translated normal to itself. The counting of N_L and $T_{A\,\mathrm{net}}$ should be restricted to those segments of the pore surfaces adjacent to grain boundaries, since these parts of the surface will determine the driving force and the area across which mass transport by volume diffusion occurs.

The linear shrinkage rate is measured for a series of samples sintered at a fixed temperature. Each sample is polished and etched for microscopic examination, with care being taken to insure that the pore edges are not distorted during this process. It is frequently advisable to impregnate with epoxy resin or a low-melting glass prior to polishing [36], although care must be taken with ceramic materials in the latter case to avoid alteration of the microstructure by the liquid glass.

The shrinkage rate and the geometric parameters at the end of each run in the series are used to compute the left-hand side of equation (11) and S_V/L_V. Figure 2 shows an example of the plots obtained by this technique [37]. The ratio of the second to the first term of the right-hand side of equation (11) is the ratio of grain-boundary to volume-diffusion flux.

Since the microstructural parameters and shrinkage rate are measured, in effect, simultaneously, complicating factors such as grain growth, vapor transport, and surface diffusion do not invalidate the use of equation (11). Moreover, the microstructural data alone can be quite useful. Correlations between S_V and porosity, for instance, can reveal surface-diffusion and/or vapor-transport contributions to sintering [37]. As more data become available, it may be possible to develop predictive sintering models.

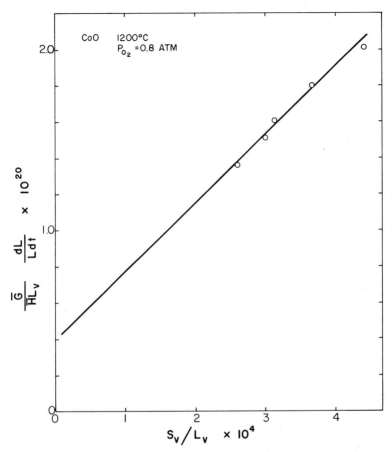

Figure 2. Intermediate-stage sintering of C_0O at 1200°C in 0.8 atm oxygen.

Conclusion

If these techniques are all employed on the same material, there will be little doubt as to the mechanism of mass transport during sintering. Various of the methods can be used by themselves to get individual parts of the total picture. The most difficult experiment is the initial-stage isothermal experiment involving spherical particles, in that uniform-sized spherical particles are rarely obtainable and often must be manufactured. The most information is obtained with a reasonable amount of effort from the intermediate-stage model. In fact, it is this model which may well provide the most useful link between fundamental studies of sintering and the practical aspects of the problem. It has very real advantages in that a detailed description of the evolution of the microstructure results from the measurements, and it is in the regime of interest in practical powder-fabrication practices.

References

1. Kuczynski, G.C., "Self-Diffusion in Sintering of Metallic Particles," *Trans. AIME*, 185 (1949), 169.
2. Kuczynski, G.C., "Measurement of Self-Diffusion of Silver Without Radioactive Tracers," *J. Appl. Phys.*, 21 (1950), 632.
3. Kingery, W.D. and Berg, M., "Study of Initial Stages of Sintering Solids by Viscous Flow, Evaporation-Condensation, and Self-Diffusion," *J. Appl. Phys.*, 26 (1955), 1205.
4. Thümmler, F. and Thomma, W., "The Sintering Process," *Met. Revs.*, no. 115, in *Metals Mater.*, 1, no. 6 (1967), 197.
5. Johnson, D.L. and Cutler, I.B., "Diffusion Sintering: I—Initial Stage Sintering Models and Their Application to Shrinkage of Powder Compacts," *J. Am. Ceram. Soc.*, 46 (1963), 541.
6. Lay, K.W. and Carter, R.E., "Time and Length Corrections in the Analysis of the Initial Stages of Diffusion-Controlled Sintering," *J. Am. Ceram. Soc.*, 52 (1969), 189.
7. Johnson, D.L., "Discussion of the 'Time and Length Corrections in the Analysis of the Initial Stages of Diffusion-Controlled Sintering,'" *J. Am. Ceram. Soc.*, 52 (1969), 562.
8. Coble, R.L., "Sintering Crystalline Solids: I," *J. Appl. Phys.*, 32 (1961), 787; and "Intermediate-Stage Sintering: Modification and Correction of a Lattice-Diffusion Model," *J. Appl. Phys.*, 36 (1965), 2327.
9. Gupta, T.K. and Coble, R.L., "Intermediate Stage Sintering: II—Experimental Data," *Sintering and Related Phenomena*, G.C. Kuczynski, N.A. Hooton, and C.F. Gibbon, eds., New York: Gordon and Breach, Science Publishers (1967), 423.
10. Kakar, A.K., "Sintering Kinetics Based on Geometric Models," *J. Am. Ceram. Soc.*, 51 (1968), 236.
11. Gupta, T.K., "Comments on 'Sintering Kinetics Based on Geometric Models,'" *J. Am. Ceram. Soc.*, 52 (1969), 166.
12. Tikkanen, M.H. and Makipirtti, S.A., "A New Phenomenological Sintering Equation," *Int. J. Powder Met.*, 1 (1965), 15.
13. Herman, M., "An Investigation of the Early Stages of Sintering," Unpublished Ph.D. dissertation, University of Pennsylvania, 1965.
14. Young, W.S., Rasmussen, S.T. and Cutler, I.B., "Determination of an Effective Viscosity of Powders as a Function of Temperature," *Ultrafine-Grain Ceramics*, J.J. Burke, N.L. Reed, and V. Weiss, eds., Syracuse, N.Y.: Syracuse University Press (1970), 185.
15. Young, W.S. and Cutler, I.B., "Initial Sintering with Constant Rates of Heating," *J. Am. Ceram. Soc.*, 53 (1970), 659.
16. Bacmann, J.J. and Cizeron, G., "Dorn Method in the Study of Initial Phase of Uranium Dioxide Sintering," *J. Am. Ceram. Soc.*, 51 (1968), 209.
17. Johnson, D.L., "Sintering Kinetics for Combined Volume, Grain Boundary, and Surface Diffusion," *Phys. Sint.*, 1 (1969), B1.
18. Morgan, C.S., "Comment on 'Dorn Method in the Study of Initial Phase of Uranium Dioxide Sintering,'" *J. Am. Ceram. Soc.*, 51 (1968), 724.
19. Morgan, C.S., "Activation Energy in Sintering," *J. Am. Ceram. Soc.*, 52 (1969), 453.
20. Rhines, F.N., "A New Viewpoint on Sintering," *Plansee Proceedings*, London: Pergamon Press (1959), 38.
21. Kronsbein, J., Buteau, L.J. Jr. and DeHoff, R.T., "Measurement of Topological Param-

eters for Description of Two-Phase Structures with Special Reference to Sintering," *Trans. AIME*, 233 (1965), 1961.

22. Barrett, L.K. and Yust, C.S., "Discussion of 'Measurement of Topological Parameters for Description of Two-Phase Structures with Special Reference to Sintering,'" *Trans. AIME*, 236 (1966), 1385.

23. Johnson, D.L. and Clarke, T.M., "Grain Boundary and Volume Diffusion in the Sintering of Silver," *Acta Met.*, 12 (1964), 1173.

24. Nichols, F.A., "Theory of Sintering of Wires by Surface Diffusion," *Acta Met.*, 16 (1968), 103.

25. Salkind, M.J., Lenel, F.V. and Ansell, G.S., "The Kinetics of Low Temperature Sintering of Loose Silver Powder," *Trans. AIME*, 233 (1965), 39.

26. Burr, M.F., Lenel, F.V. and Ansell, G.S., "Influence of Pressure Upon the Sintering Kinetics of Silver," *Trans. AIME*, 239 (1967), 557.

27. Gessinger, G.H., Lenel, F.V. and Ansell, G.S., "Continuous Observation of the Sintering of Silver Particles in the Electron Microscope," *Trans. ASM*, 61 (1968), 598.

28. Gessinger, G.H., "Effect of Atmosphere on the Sintering Kinetics and the Evolution of Microstructure," International Powder Metallurgy Conference, New York, July 1970, in *Modern Developments in Powder Metallurgy*, Vol. 4, *Processes*, H.H. Hausner, ed., New York: Plenum Press (1971), 267.

29. Hirschhorn, J.S. and Bergland, J.G., "Effect of Oxygen on the Sintering Rate of Silver Compacts," *Scripta Met.*, 2 (1968), 319.

30. Seidel, B.R. and Johnson, D.L., "Initial Sintering Kinetics of Silver," *Phys. Sint.*, 3 (1971), 143.

31. Johnson, D.L., "Methods of Determining Sintering Mechanisms," International Powder Metallurgy Conference, New York, July 1970, *Modern Developments in Powder Metallurgy*, Vol. 4, *Processes*, H.H. Hausner, ed., New York: Plenum Press (1970), 189.

32. Johnson, D.L., "New Method of Obtaining Volume, Grain Boundary, and Surface Diffusion Coefficients from Sintering Data," *J. Appl. Phys.*, 40 (1969), 192.

33. Moore, D.J., "Initial Sintering Kinetics of LiF in an Inert Atmosphere," Unpublished M.S. thesis, Northwestern University, 1968.

34. Doi, A., "Diffusion in LiF During the Initial Stage of Sintering," Unpublished M.S. thesis, Northwestern University, 1970.

35. Johnson, D.L., "A General Model for the Intermediate Stage of Sintering," *J. Am. Ceram. Soc.*, 53 (1970), 574.

36. King, A.C. and Fuchs, K., "Polished Porous Materials," *J. Am. Ceram. Soc.*, 50 (1967), 328.

37. Kumar, P., "Sintering Mechanisms in Cobalt Oxide," Unpublished Ph.D. dissertation, Northwestern University, 1972.

PROCESSING OF WROUGHT PRODUCTS (A)

MODERATOR: LEWIS R. ARONIN
Army Materials and Mechanics Research Center
Watertown, Massachusetts

9. Fundamental Principles of Powder Preform Forging

H. A. KUHN
Drexel University
Philadelphia, Pennsylvania

ABSTRACT

Development and full utilization of the powder preform forging process is predicated by an understanding of the basic phenomena involved. Attention is focused on the mechanical aspects of powder preform forging by means of experimental techniques and analytical methods normally used in the study of conventional cast and wrought materials. The densification, flow, and fracture behavior of powder preform materials (iron- and aluminum-base alloys) are determined through compression tests.

Functions relating the material behavior to its continuously changing porosity during deformation are obtained which lead to the development of a plasticity theory and methods of analysis for the forging process. A criterion for design of powder preforms based on full densification without fracture is presented by means of a graphical representation of the relations between material behavior and preform forging parameters. Through this approach, optimum or required values of porosity, lubrication coefficient, temperature, and deformation rate can be determined for a given part shape and material characteristics.

Introduction

Powder preform forging has as its objectives the deformation of a sintered powder mass into a useful shape with complete densification and no defects occurring during the process. This may be accomplished in a variety of processes from simple repressing, which involves densification but no lateral flow, to closed-die forging of a complex shape from a simple preform, which involves densification coincident with a large amount of lateral flow. Each approach involves varying degrees of complexity in preform tooling, forging tooling, and lubrication. Also, the likelihood of defect formation varies from one process to another. Each process, in addition, involves a different mode of deformation which results in different forged structures and properties. All of these factors relate directly to the economics of the over-all process, and their understanding is critical to the development of powder preform forging as a viable manufacturing process.

The central problem in powder preform forging is design of the preform in terms of shape, dimensions, and density for a given material and forging conditions. This requires an understanding of the plastic flow and densification behavior of a porous material, the subject of this chapter.

Background

Plastic deformation behavior of a conventional cast and wrought material can be determined from simple axial compression tests on cylindrical samples. If the test is performed such that no friction exists at the die contact surfaces, the free cylindrical surfaces expand uniformly, as shown in Figure 1(a). The lateral

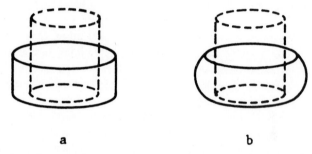

a   b

Figure 1. Lateral spread characteristics for: (a) frictionless compression; (b) compression with friction constraint.

strain is one-half the axial strain (and of opposite sign) in this case. This Poisson ratio of one-half follows from the fact that no volume change takes place during plastic deformation in a fully dense material. Also based on this fact is the theory of plasticity, expressed by the von Mises yield criterion and Levy–Mises stress–strain relations.

When axial compression tests are performed with frictional constraints at the die contact surfaces, the free cylindrical surfaces expand nonuniformly, producing a barrel shape, as depicted in Figure 1(b). The curvature of this bulging surface increases as frictional constraint increases. This mode of deformation produces tensile stresses in the circumferential direction and decreases the axial compressive stress at the equatorial surface (Figure 2). When deformation becomes large enough, surface cracks result at the bulge surface, limiting the extent of useful plastic deformation.

Sintered powder materials contain a significant amount of porosity, and their

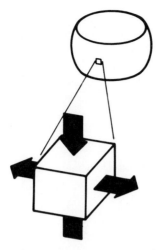

Figure 2. Stress state developed at the bulge surface of compressed cylinders with frictional constraint.

deformation behavior will show considerable deviations from that of fully dense material. These materials are subject to densification during deformation so that the associated Poisson ratio is less than one-half, and lateral spread is less than that which would occur in a fully dense material. As a further consequence, conventional plasticity theory does not apply.

When porous materials are compressed under friction conditions, the resulting lateral constraint serves the useful function of increasing densification of the material during deformation. Porosity also serves a useful function because the resulting densification during compression reduces the severity of bulge formation from that which would occur in a fully dense material undergoing the same amount of compression. This reduces tensile stresses at the bulge surface. Porosity in the material, however, reduces its effective ductility. Thus, the existence of porosity plays two counteracting roles concerning the formation of fractures at the free surface.

Present Study

The features of densification, lateral flow, and fracture described above are essential factors in the rational design of powder preforms for forging. The following presents quantitative information regarding these factors, derived from compression tests on cylindrical compacts of pressed and sintered powder. The axisymmetric geometry chosen for study simulates a large class of forged shapes; however, it is shown that the basic descriptions obtained for axisymmetric deformation apply also to plane-strain deformation.

Initial work involved the study of sintered powder preforms (iron and aluminum) at room temperature to eliminate the complicating effects of hot working. Then the same experimental methods were used in a study of hot working of an aluminum alloy powder (601 AB). The relatively low forging temperature involved in this system enabled careful control of temperature and other forging conditions. In the final section, these experimental results are utilized to establish a procedure for preform design.

Cold Deformation Study

Experimental Procedure

The basic deformation behavior of sintered powder materials was determined through room-temperature axial compression tests. One-inch-diameter cylindrical preforms of MH-100 sponge iron, Atomet 28 atomized iron, and 201 AB and 601 AB aluminum alloy powders were used. The iron powders were compacted in a free-floating die or an isostatic press with wet-bag tooling, then sintered for one hour at 2000°F in dry hydrogen. Aluminum powders were die-compacted and sintered at 1125°F in nitrogen (dew point -40°F) for one-half hour.

Tests were performed by compressing the cylinders between flat, parallel die blocks mounted on 60-ton Baldwin or 150-ton Dake hydraulic presses. Free-expansion (frictionless) tests were performed by using Teflon sheets (0.005-inch thick) at the contact surfaces. These tests involved increments of ~5-percent reduction, with the Teflon sheets replaced after each increment. Tests with friction utilized rough dies (~3,000μ finish), polished dies (~10μ), and polished dies with MoS$_2$ in oil lubricant.

Lateral Flow

Lateral flow behavior of the porous materials was determined through frictionless compression tests. This produces homogeneous axial compression without complications due to friction and permits measurement of the Poisson ratio characteristic of the material, as described in Figure 3. A slight flaring of the cylindrical surface near the top and bottom contact faces occurred in the die-compacted samples. This is probably due to the fact that such compacts are less dense at the mid-plane than at the punch-contact surfaces, and thus expand to a greater extent near the contact surfaces than at the mid-plane. The effect was not seen in isostatically compacted samples.

SIMPLE COMPRESSION

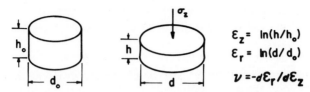

$$\varepsilon_z = \ln(h/h_o)$$
$$\varepsilon_r = \ln(d/d_o)$$
$$\nu = -d\varepsilon_r/d\varepsilon_z$$

Figure 3. Frictionless compression test, showing determination of Poisson ratio.

Preliminary results on MH-100 sponge-iron preforms revealed a relation between Poisson ratio and density given by [1]:

$$\nu = 0.5 \, (\rho/\rho_t)^{1.92} \tag{1}$$

where

(ρ/ρ_t) = fraction of theoretical density

Results of tests on other materials satisfy the same relation, as shown in Figure 4. This relation quantitatively expresses the decreasing rate of densification (increasing Poisson ratio) of a porous material under compression as density increases, reaching the value of one-half at full density.

Figure 4 indicates that there is no apparent effect of material or compaction

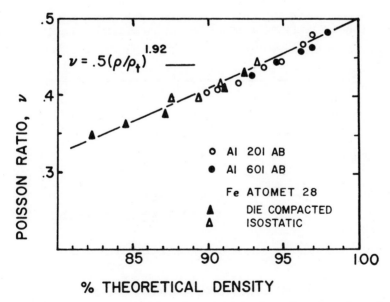

Figure 4. Relation between Poisson ratio and density for various sintered powder materials under room-temperature deformation.

method on the relation between Poisson ratio and density. A slight deviation from this line has been measured in preforms with essentially equiaxed pore shapes. (Relatively flattened pores exist in the uniaxially compressed preforms from which the data in Figure 4 is taken.) Preforms were repressed in the compaction die or the isostatic press in order to achieve deformed densities greater than the as-sintered density yet maintain an equiaxed pore shape. These samples were then tested in the usual frictionless compression manner to determine the Poisson ratio. The results, shown in Figure 5, indicate that equiaxed pores give a slight increase in the Poisson ratio. In other words, densification is less rapid, which can be readily understood from a simple model of compression of a material containing a circular hole or an elliptical hole. The latter involves more flattening of the pore, and hence more densification, for a given increment of compression.

The work-hardening exponent of the matrix material surrounding the pores is also seen to have a small effect. In all of the tested materials in Figure 4, the work-hardening exponent of the base material is between 0.2 and 0.32. In hot worked material—in which case the exponent is near zero—there is a slight reduction of the Poisson ratio, as will be seen in a later section.

Densification

Densification during frictionless compression occurs somewhat inefficiently because only the axial stress is applied, with no lateral constraining stresses. The effect of lateral constraint due to friction was measured by compressing iron

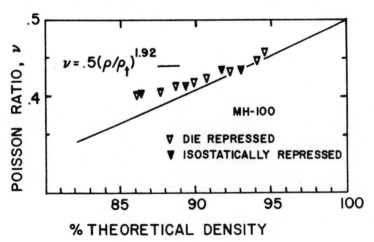

Figure 5. Relation between Poisson ratio and density for iron powder preforms predensified by repressing.

powder preforms between lubricated polished dies, unlubricated polished dies, and rough dies, in ascending order of constraint. The latter is given not for its practical value but because it closely simulates the full constraint experienced by material hot worked between smooth dies with no lubricant. Density measurements were made by machining off the bulged surface of the specimens, which are at a relatively low density, and then measuring the volume and weight of the remaining cylinder.

Results of these tests are given in Figure 6, which shows the increased densification resulting from increasing lateral constraint due to friction. Tests were

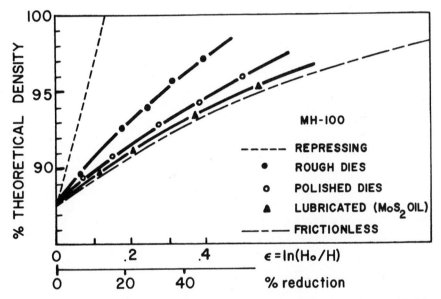

Figure 6. Density vs. axial strain for frictionless compression, compression with frictional constraint, and repressing at room temperature (MH–100 iron powder preforms).

not taken to full density because of limited press capacity. Also, fracture of the bulge surfaces occurred in every test sequence at some point prior to reaching press capacity. The cracks were not deep and were removed when the bulge was machined off, and therefore did not affect the density measurements.

For comparison, the densification curves for a repressing operation from the same preform density are included in Figure 6. Repressing, with contact at the die wall, represents the absolute maximum in lateral constraint, with no lateral flow.

Densification can be calculated in the case of frictionless compression (Figure 3) from the equation for density change

$$-d\rho/\rho = d\epsilon_r + d\epsilon_\theta + d\epsilon_z$$

and the fact that for this case

$$de_r = de_\theta = -\nu de_z$$

where

de_r = increment of radial strain
de_θ = increment of circumferential strain
de_z = increment of axial strain

giving

$$-d\rho/\rho = de_z (1 - 2\nu)$$

Using Equation (1), this becomes

$$- d\bar{\rho}/\bar{\rho}[1 - (\bar{\rho})^{1.92}] = de_z \qquad (2)$$

where

$\bar{\rho} = \rho/\rho_t$, fraction of theoretical density

Similarly, densification during repressing is given by the trivial expression

$$- d\rho/\rho = de_z \qquad (3)$$

An analysis of deformation of a cylindrical preform with frictional constraint is being developed by the author, which will enable calculation of densification during compression with frictional constraint.

Fracture

As pointed out in the previous section, fracture at the bulge surface occurred in iron powder preforms compressed with frictional constraint. This occurs because, as described in Figure 2, a tensile circumferential stress is generated at the bulge surface. The severity of this stress is directly related to the curvature of the bulge, which in turn depends on the degree of frictional constraint. Frictional constraint depends not only on the type of friction at the contact surfaces but also on the preform height-to-diameter ratio. Constraint, for a given friction condition, increases as this ratio decreases.

As an indication of the influence of friction condition and preform height-to-diameter ratio on degree of frictional constraint, and hence on surface fracture, results of compression tests to fracture are given in Figure 7, for iron powder preforms. The influence of height-to-diameter ratio and contact-surface friction condition is clear, and, at least for this system, is more significant than porosity or type of powder.

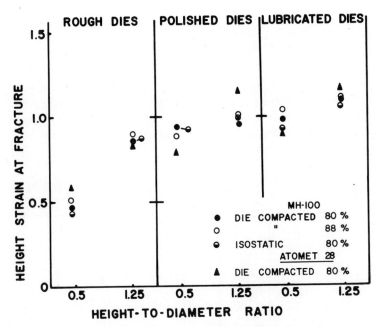

Figure 7. Height strain to fracture for iron powder preforms under room-temperature deformation.

Yield Criterion

Results of the simple, frictionless compression tests described previously suggest a yield criterion and plasticity equations for porous materials which takes into account the influence of hydrostatic pressure on yielding of such materials. The yield criterion is

$$\left[Y(\rho,\epsilon) = \left[\frac{(\sigma_1 - \sigma_2)^2 + (\sigma_2 - \sigma_3)^2 + (\sigma_3 - \sigma_1)^2}{2} + (1 - 2\nu)(\sigma_1\sigma_2 + \sigma_2\sigma_3 + \sigma_3\sigma_1) \right]^{1/2} \right.$$

$$(4)$$

where

$\sigma_1, \sigma_2, \sigma_3$ are principal stresses
ν = Poisson's ratio
$Y(\rho,\epsilon)$ = yield stress, a function of density and work-hardening

The stress-strain equation are:

$$de_1 = (d\lambda/Y) \left[\sigma_1 - \nu(\sigma_2 + \sigma_3) \right]$$
$$d\epsilon_2 = (d\lambda/Y) \left[\sigma_2 - \nu(\sigma_3 + \sigma_1) \right]$$
$$d\epsilon_3 = (d\lambda/Y) \left[\sigma_3 - \nu(\sigma_1 + \sigma_2) \right]$$

$$(5)$$

where

de_1, de_2, de_3 are incremental strains
$d\lambda$ is a proportionality constant

The applicability of these equations to the accurate prediction of forging pressures in repressing and plane-strain processes has been demonstrated previously [2]. They can also be utilized to derive a densification relation for plane-strain deformation. In plane strain, as shown in Figure 8 (replacing subscripts 1,2,3 by

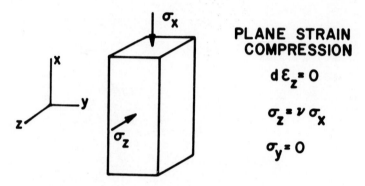

Figure 8. Stress in plane-strain deformation.

x,y,z), $de_z = 0$ and $\sigma_y = 0$, giving $\sigma_z = \nu\sigma_x$ so that

$$de_y = (d\lambda/Y) \, [- \nu(1 + \nu)\sigma_x]$$

and

$$de_x = (d\lambda/Y) \, [(1 - \nu^2)\sigma_x]$$

Then

$$d\rho/\rho = de_y + de_x = de_x [(1 - 2\nu)/(1 - \nu)]$$

and substituting equation (1), gives

$$\frac{- d\rho[1 - 0.5(\rho)^{1.92}]}{\rho[1 - (\rho)^{1.92}]} = de_z \tag{6}$$

The accuracy of this expression was tested by applying it to data from plane-strain tests presented by Antes [3]. The calculated results are given in Figure 9 for comparison with experimental data for plane-strain compression of an atomized iron powder, a 1020 steel powder, and a 1040 steel powder. The close agreement demonstrates the universal applicability of the Poisson ratio-density relation [equation (1)] and the plasticity theory [equations (4) and (5)] to cold forging of porous preforms.

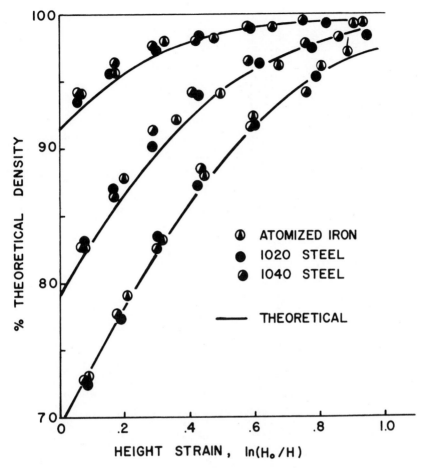

Figure 9. Comparison of experimental (Reference [3]) and theoretical [equation (6)] results for densification of iron and steel preforms by plane-strain compression at room temperature.

Hot Working Study

The generalizations regarding densification, flow, and fracture of powder preform materials given in the previous section were developed from compression tests at room temperature. In this section, the applicability of these generalizations to hot working of powder preforms is tested. To perform this study in a way in which careful control of all forging conditions could be maintained, an aluminum alloy powder 601 AB was utilized. The forging temperature was 700°F, and forging was carried out with heated dies at 625°F to minimize heat loss from the preforms. This also permitted the use of Teflon for the frictionless compression tests since 700°F is below its toxic range. The tests were carried out on a 150-ton Dake hydraulic press at an approximate strain rate of 1 sec^{-1}.

Lateral Flow

Frictionless compression tests were conducted on aluminum preforms at 700°F, using the same technique as described in the previous section for room-temperature tests. Figure 10 gives Poisson ratio versus density for the aluminum

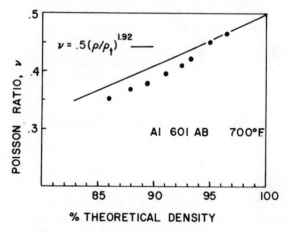

Figure 10. Relation between Poisson ratio and density for hot forging of 601 AB preforms (preform density, 83%).

preforms, along with the relationship [equation (1)] established for room-temperature deformation. It is clear that the Poisson ratio is lower in this hot working case than in room-temperature deformation. This probably results from the lack of cold working in the matrix material at elevated temperatures, which spreads the deformation more uniformly throughout the metal-pore system and gives greater densification than in a system in which the matrix is cold worked.

From the frictionless compression tests, the stress–strain relation for the hot working of aluminum preforms is given in Figure 11, along with that for cold worked aluminum preforms. For the hot worked preforms, there is a slight increase in yield stress with strain due to increasing density. Work-hardening of aluminum preforms during room-temperature deformation is due to cold working of the matrix material as well as densification.

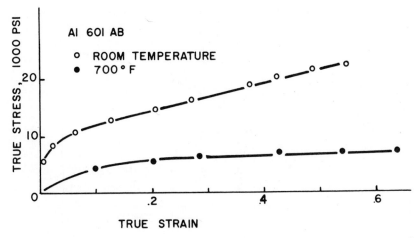

Figure 11. Stress–strain curve for 601 AB aluminum alloy powder preforms at 700°F, compared with that at room temperature.

Densification

Densification of aluminum preforms during compression under friction conditions was determined through upset tests between smooth dies and also smooth dies with MoS_2 in oil-base lubricant. Three different preform height-to-diameter ratios were used. Density was determined as before by machining off the bulge surface and measuring the weight and volume of the remaining cylinder. Results are shown in Figure 12, along with the densification curve for frictionless compression. It is evident that as the degree of frictional constraint increases (increasing friction, decreasing height-to-diameter ratio), the degree of densification for a given axial compression increases.

Fracture

As in the room-temperature tests on iron powder preforms, surface fractures occurred in the aluminum preforms hot forged with frictional constraint. The effect of frictional constraint (friction, height-to-diameter ratio) on fracture is given in Figure 13. Again, as constraint decreases (friction decreased, height-to-diameter increased), the compressive strain to fracture increases. It can be expected that this general trend will be complicated by strain-rate effects and by the use of cold dies, resulting in nonuniform temperature distributions.

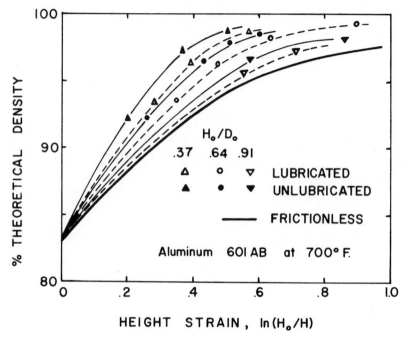

Figure 12. Densification during hot forging of 601 AB preforms.

Preform Design

Specification of the powder preform shape, dimensions, and density for successful forging of a given shape is a key problem in the over-all process. It is not possible to give general guidelines for preform design at the present time; however, densification curves such as those given in Figures 6 and 12 can be used to predict the amount of strain required for full densification under specific friction conditions. Although this information is the result of axisymmetric compression tests, the results can nevertheless be applied to multi-level and complex parts by considering the densification of each individual section of the part. Generally, in order to prevent folds or cracks between any two sections, it is best to distribute metal in the preform such that flow from one section to another is not required.

Preform design will not simply involve flow and densification considerations in some material systems. As in the cold forged iron and hot forged aluminum alloys considered in this work, fracture occurs before full density is reached. Design of the preform in these cases must be such that the expanding surfaces of the deforming preform reach the die sidewalls before surface fracture occurs. Then the additional constraint supplied by the die sidewalls will prevent surface

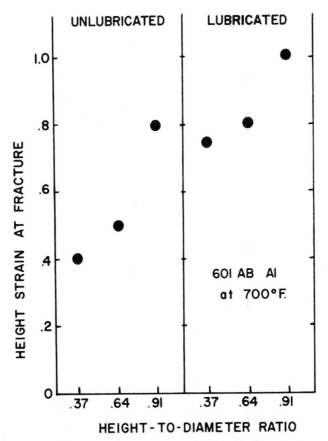

Figure 13. Height strain at fracture for hot forged 601 AB aluminum alloy preforms (83% initial density).

fractures from forming and remaining densification will be achieved in a repressing mode.

Consideration of these factors (densification, compression to fracture limit, and repressing) can be represented graphically, as shown in Figure 14. In this illustration, the axes represent preform and forged part geometries; the ordinate is preform height-to-diameter ratio and the abscissa is the forged height-to-diameter ratio. A forging operation can be traced by starting at the preform H_0/D_0 along line OA. As forging commences, the forged H/D decreases, tracing a horizontal line to the left. Eventually, after some degree of deformation determined by friction and geometry parameters, either full density is reached, after which the process can cease, or fracture is imminent. In the latter case, the expanding surfaces of the material must meet the die sidewalls such that repressing

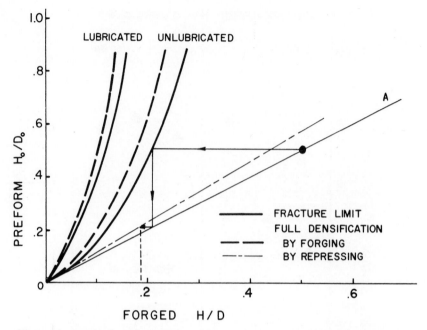

Figure 14. Graphical representation of fracture and densification behavior of porous materials for preform design.

occurs, preventing fracture and closing up the last remaining porosity. Retracing these steps, for a required forged H/D, the amount of repressing necessary is first found such that fracture is avoided. This, coupled with the fracture line, defines the maximum preform H_0/D_0 for successful forging.

The fracture curves and full-density curves given in Figure 14 are taken from data presented in the previous section for hot forging 601 AB aluminum alloy powder. The fracture curves are derived from Figure 13 and the densification curves are estimated from Figure 12.

Both curves swing to the left as friction is reduced. In the limiting case of zero friction, both curves coincide with the ordinate. That is, the homogeneous deformation resulting from frictionless compression will not cause fracture, but full density also cannot be achieved.

The information given in Figure 14 is for the single density studied (83 percent). It is apparent, however, that higher densities will swing the densification curves to the right and the fracture curve to the left. Additional testing is necessary to determine the magnitude of this movement for a given increase in preform density.

Summary

It has been demonstrated that the densification, flow, and fracture behavior of a hot worked powder preform material closely resembles that for cold forged materials. Similar methods, namely the upset test, may therefore be used to determine hot working behavior of a given material.

The influence of frictional constraint on densification during forging is given, showing that decreasing preform height-to-diameter ratio and increasing friction increase densification for a given amount of compression. These same factors, however, decrease the amount of allowable compression to fracture.

Preform design can be approached graphically by considering allowable deformation to fracture and then accomplishing the remaining densification by repressing.

Acknowledgment

The author gratefully acknowledges the contributions of Mr. C. L. Downey and Mr. R. P. Brobst for the experimental work reported here. Appreciation is also extended to Mr. James Dowd of the Alcoa Technical Center for his contribution of the aluminum powders and sintered aluminum compacts. This research is sponsored by the Department of Defense under a THEMIS project (Technical Liaison with the Frankford Arsenal, Philadelphia, Pennsylvania).

References

1. Kuhn, H.A., Hagerty, M.M., Gaigher, H.L. and Lawley, A., "Deformation Characteristics of Iron-Powder Compacts," International Powder Metallurgy Conference, New York, July 1970, in *Modern Developments in Powder Metallurgy*, Vol. 4, *Processes*, H.H. Hausner, ed., New York: Plenum Press (1971), 463.
2. Kuhn, H.A. and Downey, C.L., "Deformation Characteristics and Plasticity Theory of Sintered Powder Materials," *Int. J. Powder Met.*, 7 (1971), 15.
3. Antes, H.W., "Cold and Hot Forging P/M Preforms," SME International Engineering Conference, Philadelphia, Pennsylvania, 1971.

10. Processing and Properties of Powder Forgings

H. W. ANTES
Hoeganaes Corporation
Riverton, New Jersey

ABSTRACT

Three distinct powder-metallurgy (P/M) forging processes have evolved. Each of the processes involve three stages: preform manufacture, forging, and finishing. The P/M forging processes identified in terms of the forging stage are: (1) hot repressing, (2) confined-die flashless, and (3) closed-die limited flash. The importance of preform design and proper preform processing are emphasized. Preform density and geometry are covered. Examples of preforms with accompanying forgings are presented. Techniques for preform heating are covered and the importance of adequate protection against oxidation is stressed. A comparison is made between simple types of deformation and the deformations experienced in the actual P/M forging processes. Deformation-densification data are presented with a method for calculating densification from strain measurements. Forgeability is discussed in terms of the results of high-temperature torsion tests of P/M preforms. The effect of oxidation on forgeability is also covered. The hardness, tensile, impact, and fatigue properties of P/M forgings are presented with a discussion of the effect of processing variables on mechanical properties.

Introduction

Conventional P/M (compacting and sintering) has been used for many years to make structural components. High-strength components have been made through the use of heat-treatable materials, and by special treatments such as double-pressing and sintering or infiltration to reduce the amount of residual porosity. However, components made by these processes have had mechanical property limitations with respect to impact or fatigue strength, and even ductility. These limitations may be directly attributed to the residual porosity. If maximum properties are to be realized from components made of powder, porosity must be eliminated or at least significantly reduced over that which is realized through conventional pressing and sintering techniques. Forging powder-

171

metal preforms is a process that can be used to reduce and even completely eliminate this porosity.

It is the purpose of this chapter to describe the various P/M forging processes that have evolved and the variables which must be considered in these processes. Furthermore, some of the fundamentals of the processes will be discussed and the special considerations necessary above those for conventional forging processes will be presented. Mechanical properties will be presented and processing factors that affect these properties will be described. Most of the data will be for iron and low-alloy steels. However, most of the principles presented may be applied to virtually any powder metal or alloy.

Material Selection

In the case of ferrous materials, there are three principal types of powders available. The types are: (1) electrolytic, (2) sponge, and (3) atomized. Photomicrographs of these three types of powder are shown in Figure 1. Electrolytic powder is very high-purity material, but it is expensive and will probably have limited use in P/M forging because of its high cost. Furthermore, the powder particles are somewhat rounded and the green strength of this material is low. Sponge iron has an irregular type of particle and has good green strength even at low green density, which may be important for complex preform shapes. Sponge iron is inexpensive, but alloying must be achieved for the most part by mixing elemental powders. Furthermore, sponge irons may contain anywhere from 1 to 3 percent of insoluble second-phase material. This may not be undesirable if less than full density is required in the forging. However, if full density is required and if maximum properties are to be achieved, then an atomized type of powder will probably be preferred. Atomized powder particles have a shape somewhere between that of sponge and electrolytic powders. They have good green strength and high compressibility. Atomized powder can be made in a wide variety of alloy compositions by prealloying in the melt so that each powder particle is completely alloyed. These factors, coupled with the low cost and the relative cleanliness of atomized powder, dictate using an atomized product for most P/M forging applications.

Two approaches have been taken to achieve the desired alloy compositions in steels. These are: (1) direct atomization of alloy steel (with or without carbon), and (2) mixing iron powder (atomized or other types) with powders of other elements or compounds. In the second case, some alloying is accomplished via diffusion during sintering. In the first case (atomized alloy powder), a high degree of homogeneity can be achieved since each powder particle is fully alloyed. Alloy powder containing carbon has the advantage (over those that must use graphite addition to achieve the desired carbon content) that lower sintering

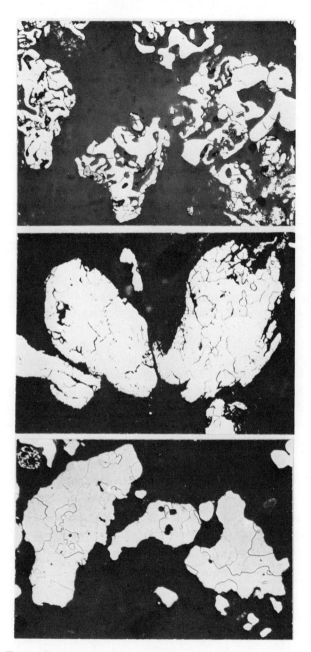

Figure 1. Types of iron powders. Top, sponge; center, electrolytic; bottom, atomized.

temperatures can be used. Preforms made from mixtures of alloy powder with graphite must be sintered at higher temperatures to efficiently adsorb and diffuse the carbon. One disadvantage of using the precarburized powders is that higher loads are required in compacting to achieve a given preform density. The desirability of low or high preform density will vary with the P/M forging process being used and the finished shape of the forging. This factor will be discussed in more detail later.

Powder Forging Processes

The basic P/M forging process involves the following four steps: (1) selection of metal powder, (2) fabrication of preform, (3) forge preform, and (4) finishing operations on forging.

Three variations of the basic P/M forging process have evolved. The processes are illustrated diagramatically in Figure 2. The process on the left utilizes a precision preform made by die-pressing to a shape very similar to the final forged configuration. Therefore, the geometry and configuration of the preform is fairly well defined in this process. The forging step in this process is more accurately described by the term "hot repressing."

The second or center process utilizes flashless forging. In this process, the preform can vary from something relatively easy to design to one probably the most difficult to design. In flashless forging, the preform weight must be controlled very accurately. If the preform is too light, undersized or less-than-full-density forgings will result. If the preform is too heavy, the forging may be oversized or die overload or die failure may occur. In one automated forging process the preform weight was controlled to ±0.5 percent [1].

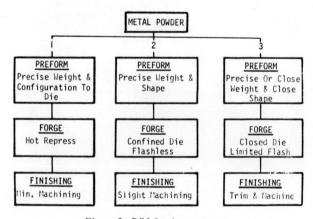

Figure 2. P/M forging processes.

The third process (right-hand side of Figure 2), utilizes conventional closed-die forging techniques. The weight of the preform used in this process does not have to be as carefully controlled as that in flashless forging. The weight variation can be compensated by the amount of flash produced.

The three processes given above represent those processes used extensively for powder forging. The first process, "hot repress," has been used predominantly for components where full density of the part is not required. Residual porosity of the forging may range from almost zero to 2 percent voids. Hot repressing pressures as high as 96.8 tsi have been reported [2]. The flashless and conventional closed-die forging processes are used when full density is required. However, powder forgings made by these processes are not always completely fully dense.

The hot repressing process has been used principally by people with a P/M background. The closed-die limited flash process has been used principally by people with a forging background. It is important in this process that the preform is hit only once (or as few times as possible) in one die impression, as opposed to conventional forging where multiple die impressions and multiple blows in each impression are used to produce the final forging. The confined-die flashless process is somewhat of a combination of the other two processes. In this process, the forging is made with a single blow or press on the preform, and lateral flow occurs during compaction but no flash is formed because confined or "trap" dies are used.

The individual P/M forging process chosen to manufacture a part will depend on the particular component and the company which will make it. It appears that each process will have its place in contributing to establishing a P/M forging industry.

Preforms

Preform Designs

Before going into the fundamentals of preform design, let us examine some of the preforms that have been used and then reflect back on these designs later. A powder-forged bearing is shown in Figure 3. The preform is shown on the left and the forging on the right [3]. The "hot repress" type of process (illustrated on the left in Figure 2) was used to make this part. The similarity between the geometry of the preform and final forging is characteristic of this process. Another example of a preform used in a hot repress type of forging process is shown in Figure 4. The preform for the gun hammer is shown in the center of Figure 4. The final forging is shown to the right and left of the preform [4]. Notice once

Figure 3. Hot repressed powder preform (left) and bearing forging (right) [3].

again the typical similarity in geometry of the preform and forging for the hot repress type of process.

Examples of preforms used for flashless forging in a confined die (the center process in Figure 2) are shown in Figures 5 and 6. The preform for the well-

Figure 4. Hot repressed powder preform (center) and gun-hammer forgings (left and right) [4].

Figure 5. Confined-die forging preform (left) and forged pinion gear (right).

known pinion gear is shown on the right of Figure 5, with the forging on the left. Here we see a preform with relatively simple geometry and a significantly different shape than the final forging. It can be seen from this figure that during forging there is appreciable lateral flow of material to produce the teeth in this gear. In the case of the connecting rod shown in Figure 6, the preforms are shown on the bottom of a and b, with the forging on the top. At first glance, this connecting rod appears to be a hot repressing. However, considerable metal flow has taken place in the shank portion. The much higher degree of symmetry of the pinion gear required a much less critically designed preform than the connecting rod. In general, as the symmetry of the final forged shape decreases, the more critical the design of the preform becomes.

The third process shown in Figure 2 (on the right side) utilized conventional closed-die forging where flash is produced. A link that was made using this technique is shown in Figure 7, with the preform. The preform is shown on the left,

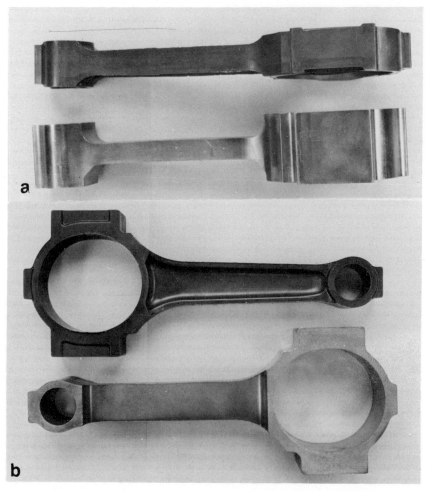

Figure 6. Confined-die powder preform [bottom in (a) and (b)] and forged connecting
rod [top in (a) and (b)] .

with as-forged link in the center and the forged link with the flash trimmed is
shown on the right [5] .

All of the preforms shown in Figures 3–7 were made by compacting in die
sets. However, other techniques are available for compacting powders into preforms for forging. Isostatic compaction is one of these techniques that has been
used successfully and has potential for commercial application.

Theoretical and Practical Considerations for Preforms

In the design of a preform, the one fundamental starting point is the known
weight of the forging (plus the weight of the flash if it is produced in the pro-

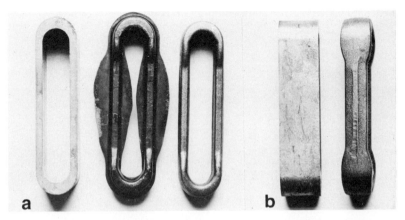

Figure 7. Closed-die powder preform [left in (a) and (b)], forged link with flash [center in (a)], and forged link with flash removed [right in (a) and (b)].

cess). The problem then is what the preform size (density) and shape (degree and type of flow) should be to facilitate making the forging. Unfortunately, the fundamental information that would be required for formal design criteria has either not been completely developed or is not being divulged. Ideally, if preform densification data as a function of straining were developed, then preforms could be designed that would provide the strain required during deformation to yield the required density. Some of these data have been published for the cold deformation of atomized iron and steel preforms and are shown in Figure 8 [6]. These data relate the preform density to final density for the three types of deformation shown in Figure 9. The uniaxial compression type of deformation is a simple upset that occurs in the first stage of confined-die flashless forging and conventional closed-die forging. This type of deformation is also experienced in upsetting of billets in open-die forging. The plane-strain type of deformation is simulated in closed- and confined-die forging when the preform begins to touch the side of the forging die and lateral flow is partially restricted. The repress type of deformation occurs when the die cavity is filled immediately before closure, or actually from the beginning of the forging in the hot repress type of forging process. The data in Figure 8 are for atomized iron preforms; similar data for low-alloy steel atomized-powder preforms are given in Figure 10. These densification curves represent the behavior of iron-carbon alloys (up to 0.4 percent carbon) and other low-alloy steels (8620, 4620, 4640). The practical aspect of the fact that a single set of curves represent the behavior of several alloy steels implies that development of a preform design is not sensitive to alloy composition. It is data of this type that should facilitate preform design; however, since sufficient data of this type are not available, there are other guidelines that may be followed in preform design. These guides may be stated as follows for closed- or confined-die type forgings:

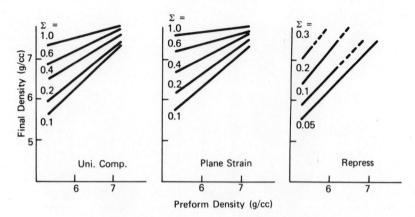

Figure 8. Densification—atomized Fe preform.

1. Design for nonplanar contact between preform and punches or top and bottom die surfaces. This minimizes the interfacial friction and promotes metal flow (smaller dead-metal zones) and decreases surface porosity.

2. Lower density preforms tend to have better filling capacity into thin or fine details [7].

3. Higher density preforms tend to require somewhat lower loads to achieve a given final fixed density for both cold and hot forging [6,8].

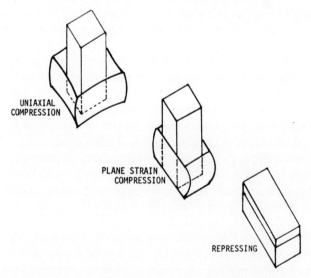

Figure 9. Types of deformation used in densification studies.

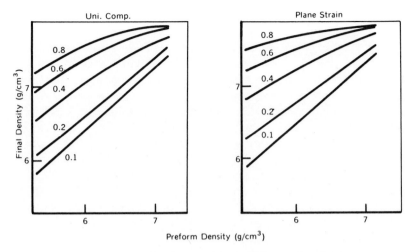

Figure 10. Densification curves for low-alloy steel preforms.

4. In order to minimize cracking during extrusion in the forging, flow should take place into a continuing decreasing cross-sectional area.

5. Straight-walled protuberance may be produced, provided that a retracting-type punch that supplies some compression stresses on the leading surface is used.

6. Increasing the degree of sintering tends to increase the amount of deformation or free flow possible before cracking or fracture of the preform.

A good example of item 1 (nonplanar contact between preform and punch) is illustrated in Figure 5 for the pinion gear. The preform is flat on the top, while the forging is curved. The concave tooling coupled with the flat preform provided good flow and compressive stresses during the deformation. Compressive-type stresses are important in any deformation process because they contribute to an apparent increase in formability (deformation before fracture).

The density of preforms that have been used vary over an extremely wide range. Many applications utilized preforms in the range of 6.2–6.8 g/cc. However, there have been cases where the preform density was less than 6.0 g/cc and other cases where the density was greater than 7.3 g/cc.

The problem of minimizing cracking in lateral flow or extrusion, either with a pressure-loaded retracting punch or a decreasing cross section of area cavity, has not been completely solved. However, these techniques are used in certain instances.

Preform Fabrication

There are a variety of techniques that can be used for compacting powder into preform shapes. These include mechanical or hydraulic compacting in either a hot or cold die, isostatic compaction, explosive compaction, slip casting, and loose powder sintering. However, the principal techniques used for making iron and steel P/M forging preforms are cold die-compacting and isostatic compacting. The precision required for the preforms used in the hot repress type of process dictates that die-compaction be used for making the preforms. The flashless forging process and the conventional forging process may use preforms made by either die-compaction or isostatic compaction. The decision as to which method to use for making the preform depends on many factors, such as number or required parts, type of equipment available, size of the forging, and the forging process as well.

High volume of a part that requires a preform of 15 pounds or less dictates a die-compaction technique be selected for making preforms. Then a decision must be made concerning whether lubricant should be mixed in the powder or a die-wall type of lubricant system be used. Mixing the lubricant with the powder requires a burn-off of the lubricant. Improper burn-off can cause rupture of the preform, especially in larger and/or high-density preforms. Other problems are incomplete burn-off or undesirable residues left after burn-off. Die-wall lubrication virtually eliminates these problems. However, die-wall lubrication is somewhat limited to simple shapes. Interestingly, the geometry of many powder-forging preforms are simple enough so that die-wall lubricating could be used.

A low volume of required parts or parts requiring preforms larger than 15 pounds may dictate that isostatic compaction techniques be used for making the preforms. Lubricants are not used in isostatic compaction techniques; therefore, no burn-off is required. The two techniques used in isostatic compaction are: (1) wet-bag, and (2) dry-bag.

The wet-bag process consists of filling a rubber bag shaped to the desired geometry with powder, plugging the end of the bag, immersing the assembly into the fluid of an isostatic compaction chamber, and pumping the chamber up to the desired pressure. The pressure is released, the assembly removed from the fluid, and the bag stripped from the preform. A wooden pattern, bag, and the connecting-rod preform made using this technique are shown in Figure 11 [9].

The dry-bag process utilizes a polyurathane split mold such as that shown in Figure 12. In this process the mold is assembled, powder poured into the mold, and the assembly then placed in an isostatic compaction chamber that is fitted with a rubber membrane which separates the fluid from the split mold. The fluid is pumped to the desired pressure, then pressure is released and the preform removed by opening the mold. Automated techniques have been developed for the dry process, and substantial numbers of preforms can be made in a unit time. The actual numbers depend on the size of the preform and the system used.

Sintering of compacted preforms is currently being used as common practice prior to the forging operation. During sintering, it is necessary to maintain the proper atmosphere that will provide protection against oxidation and proper con-

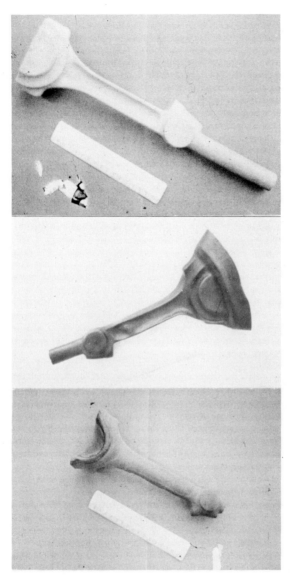

Figure 11. Isostatic compaction wet-bag. Top, wooden pattern; center, rubber bag; bottom, powder preform.

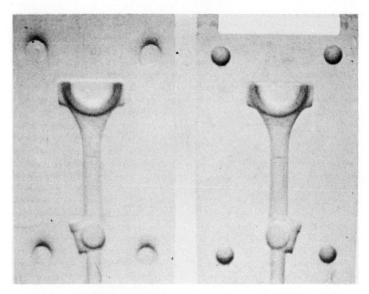

Figure 12. Polyurethane split mold used in isostatic dry-bag process.

trol of carbon in the preform. One of the purposes of sintering preforms is to provide the desired combined carbon from graphite additions made to the iron or steel powder. In addition, elemental powders of other elements added can diffuse into solution while the interparticle bond strength increases during sintering. If a completely alloyed powder is used to make the preform, then the temperature of sintering may be significantly reduced or the sintering operation may be completely eliminated and the green preform heated directly to the forging temperature and then forged.

Other approaches to preform processing include induction-heating of the green preform for sintering. Although induction sintering has not been fully developed, it has been reported that temperatures of 2100°F and above are required to attain good strength and carbon solution in compacts with blended graphite. Furthermore, unsintered compacts induction-heated to 2100°F and hot forged exhibited the same strengths and densities as those forgings made from compacts that had been furnace-sintered for 30–40 minutes at 2050°F [10].

High-Temperature Oxidation and Decarburization of Preforms

Preforms may be hot forged directly from the sintering furnace, cooled to some intermediate temperature lower than the sintering temperature, and then forged or cooled to room temperature and then reheated to the forging temperature. Whatever the process, the preform is usually exposed to air for some small but finite period of time prior to forging. During this exposure, the preform can deteriorate by oxidizing both on the exterior and interior or by decarburizing. The type and degree of preform degradation depends on temperature, preform density, and duration of exposure. Published results of a recent study show the effect of these variables on the degradation of preforms made of 1040-type material [11]. Some of these data are presented in Figure 13. These data indicated that even at these relatively low forging temperatures the tolerable exposure time is relatively short. The tolerable exposure times before internal oxidation occurred in preforms made of 1040-type material are shown in Figure 14. It can be seen from this figure that the tolerable exposure time increases with increasing preform density. However, the tolerable exposure time is a more complex function of temperature. At lower temperatures (1100–1500°F) it is relatively short. It increases to maximum times at about 1800°F and then decreases to very short times at 2050°F. This behavior

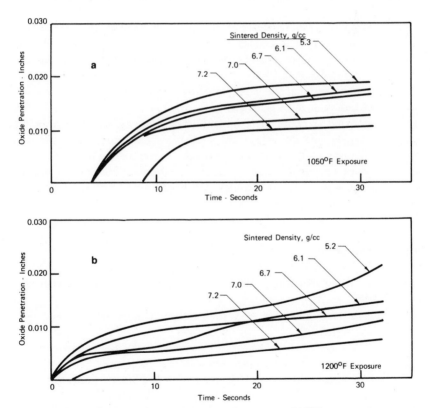

Figure 13. Oxide penetration of exposed preforms. (a) 1050°F exposure temperature; (b) 1200°F exposure temperature.

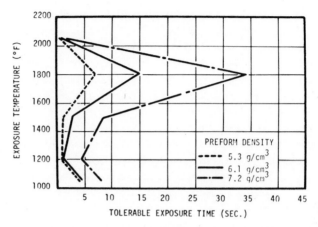

Figure 14. Tolerable exposure time for 1040-type preforms heated for forging in endogas atmosphere.

can be explained thermodynamically. Since the material is essentially iron plus carbon, only two reactions are possible when the preforms are exposed to air—iron can oxidize or carbon can oxidize. At lower temperatures, the iron oxide reaction is thermodynamically more favorable and the preform oxidizes readily. As the temperature is increased, it becomes more favorable for the carbon to combine with oxygen and a sacrificial type of decarburization takes place, delaying the iron oxide reaction. At the highest temperatures, decarburization takes place very rapidly and the delay time for the iron oxide reaction is almost nil; therefore, the iron oxidizes rapidly. However, by precoating the preforms with various media prior to preheating, all tolerable exposure times may be extended. Resin-base and water-graphite coatings have been found to be successful in this respect.

Preform Summary

In designing and fabricating P/M preforms for forging, the principal considerations include:

1. *Selection of forging process*—(a) hot repress, (b) flashless confined-die, (c) conventional closed-die;

2. *Material selection*—(a) prealloyed atomized powder, (b) premixed powders (atomized or other), (c) reduced powders;

3. *Preform design*—(a) prior art, (b) theoretical and practical design criteria;

4. *Preform fabrication*—(a) type of compaction, (b) sintering;

5. *Effects of exposure to air prior to forging*—(a) natural tolerable exposure time, (b) methods for increasing tolerable exposure time.

The considerations are extensive and complex in some cases. However, the integrity of the powder forging can be no better than the preform from which it was made.

Deformation-Densification

Fundamentals of Deformation

In classical plasticity theory, the first principal assumptions are that a material is continuous, homogeneous, and isotropic, and that no volume change occurs during plastic deformation. Conventional fully dense, fine-grain size, cubic lattice materials approach these requirements, and yielding and deformation behavior can be predicted from various theories. As an example, yielding can be predicted fairly accurately for any general type of loading by the von Mises or or Distortion Energy Theory [12]. According to these theories, yielding and flow are not affected by the hydrostatic component of the stress system owing to the fact the hydrostatic component is responsible for volume changes only. However, any material that has a large amount of internal porosity should yield under the influence of hydrostatic-type loading because a significant inelastic volume change can be achieved. Therefore, theories of yielding and flow of P/M preforms must be an extension of current classical flow theories. A theory of flow for cold deformation of P/M preforms has been presented by Kuhn and Downey [13], who consider the effect of the hydrostatic stress on yielding. However, for the purpose of simplication in this chapter, a qualitative approach will be taken.

The problem involved in P/M forging may be resolved into forming the desired shape while at the same time closing voids and deforming the powder preform in such a way as to minimize degrading affects. The mechanism for the closure of voids must be one involving shear flow. Consider a void indicated by the hatched area in the body shown in Figure 15(a). If only a hydrostatic stress is applied to the body, as in Figure 15(b), the volume of the void may be made to decrease through elastic and plastic deformation, but an infinite pressure would be required to completely eliminate the void by hydrostatic loading [14]. One mechanism whereby voids could be closed is illustrated in Figures 15(c) and (d). In this process, a shearing stress is superimposed on the hydrostatic stress. The void folds over and closes during the deformation. This shearing deformation occurs in all three types of the P/M forging processes shown in Figure 2. However, the amount of shear flow is smallest in the hot repressing process.

The problem of what may happen to inclusions during deformation is illustrated in Figure 16. The inclusions are represented by the hatched area within an undeformed matrix in Figure 16(a). If the relative strength of the inclusion

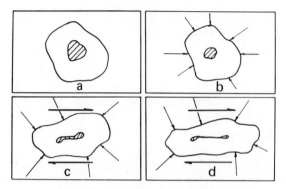

Figure 15. Deformation of voids.

to the matrix is high, the inclusion may not deform and cracks or voids may open at the matrix inclusion interface, as illustrated by the black regions in Figure 16 (b).   One factor that helps to minimize void formation is a high hydrostatic stress.   However, if the inclusion is not significantly stronger than the matrix, then the inclusion may deform and be elongated. There is a tendency for voids to be formed at the ends of deformed inclusions as well [Figure 16(c)]. Increasing the deformation beyond the state shown in (c) may cause the inclusion to fracture and leave voids between the fragments [Figure 16(d)]. These phenomena have been observed in P/M forging [15].

Types of Deformation

The type of deformation affects the rate of densification. The three simple types of deformation (illustrated in Figure 9) that, singly or combined, represent the P/M forging processes are: (a) uniaxial strain (repressing), (b) biaxial strain (plane-strain compression), and (c) triaxial strain (uniaxial compression). These processes are listed in decreasing rate of densification per unit amount of compaction strain. However, the forging pressure or true stress required to densify a P/M preform to some fixed value is greatest for repressing and least for uniaxial compression. This is because flow facilitates densification and, in repressing, the only freedom for flow is achieved through reduction of the internal porosity. Therefore, parts made by repressing usually have up to about 2 percent residual porosity. The other two forging processes, confined-die flashless and closed-die with limited flash, have external freedom for flow and all three of the above types of deformation occur. During forging in the external flow processes, the initial stages of preform deformation are similar to uniaxial compression. As the deforming preform touches parts of the die wall, some restraint to flow is imposed and then the deformation is similar to plane-strain compression. Finally,

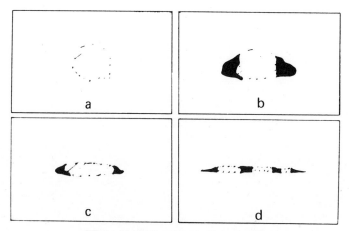

Figure 16. Deformation of inclusions.

immediately before total closure, the preform essentially fills the die and the last step of the process is a repressing.

The total amount of flow is important because it may affect the magnitude and anisotropy of properties. Large amounts of flow facilitate pore closure, but tend to open voids around inclusions unless the hydrostatic component of the stress is high. Small amounts of flow will minimize mechanical fibering and thus minimize mechanical property anisotropy. However, porosity may not be closed up completely.

The temperature of the forging process is important because the ease of densification increases with increasing temperature. Closely related to temperature is the speed of the forging process. The faster the forging process, the less the part cools in the die. Therefore, the better the densification and the lower the forging pressure.

Lubrication

Proper lubrication for either hot or cold forging is important because it extends die life and better densification of the surface is achieved. Furthermore, proper lubrication can reduce forging loads and promote more homogeneous deformation. In the case of hot forging, the lubricant provides thermal insulation between workpiece and the die-surface finish.

In the case of hot forging, the lubricant used most is a colloidal dispersion of graphite in water. Nonuniform lubrication or nonuniform temperature in upper and lower punches or dies can contribute to nonsymetrical deformation. In Figure 17, a $2'' \times 2'' \times 5''$ preform is shown in the center with hot forgings to the

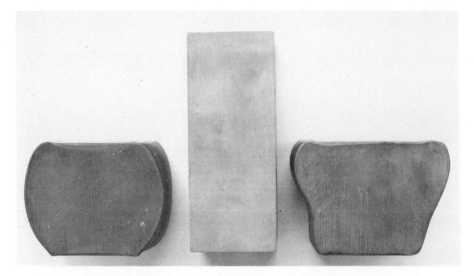

Figure 17. Plane-strain hot forgings with preform.   Left, forging with uniform friction top and bottom; center, preform; right, forging, lower friction at the top.

left and right. Forging was accomplished by plane-strain upsetting. The forging on the left had similar friction at the top and bottom. The forging on the right had lower friction at the top and a higher friction on the bottom.

Lubricants that have been used for cold forging include Teflon and graphite.

Densification

It is interesting to note that in Figure 10 a single set of curves represent the densification behavior of several alloy steels. These curves also represent the deformation-densification of carbon steel preforms in the range of 0-0.40 percent carbon. These results indicate that a common denominator for relating densification for various types of materials and geometries may be *strain*, or more precisely the sum of the normal true strains.

Since, $\epsilon_x + \epsilon_y + \epsilon_z = \Delta$ (change in volume/unit volume), then densification may be calculated from the strains. For the case of plane-strain deformation, $\epsilon_z = 0$. Therefore, $\Delta = \epsilon_x + \epsilon_y$. Plane-strain compression tests were made on $1\frac{1}{4}'' \times \frac{1}{2}'' \times \frac{1}{2}''$ P/M preforms. Compaction was in the $1\frac{1}{4}''$ (x) direction with lateral flow in one of the $\frac{1}{2}''$ (y) directions. The strains were calculated by:

$$\epsilon_x = 1n \frac{x_f}{x_0}$$

$$\epsilon_y = \ln \frac{A_z/x_f}{y_0}$$

Where

$x_0 = 1\frac{1}{4}''$
A_z = area of z face
A_z/x_f = average deformed length in y direction
$y_0 = \frac{1}{2}''$

Δ, was determined from:

$$\Delta = \epsilon_x + \epsilon_y$$

Density was then calculated as follows:

$$\text{Total change in volume} = V_0 \Delta$$

$$\text{Calculated density} = \frac{\text{Wt}}{V_0 - V_0 \Delta}$$

Where

V_0 = original volume of preform
Wt = weight of preform

Measured density was determined by using an Archimedean technique. Calculated and measured densities are compared in Table I.

The data in Table I indicate that densities calculated from deformation strains agree well with actual measured densities. Although the range in variation in the differences in densities were approximately 5–6 percent maximum, most of the data were in agreement to within 2–3 percent.

This implies that at least for simple shapes (*e.g.*, rectangular prisms, cylinders, etc.), preform density and dimensions may be calculated so that in closed-die (trap-die) flashless forging, essentially full density can be achieved when the preform flows to the die walls. This minimizes the amount of repressing required in the final steps of deformation.

Although the results shown here are for simple types of deformation and simple geometries, this work is being extended and laboratory tests are in progress that will contribute to furthering the understanding of deformation-densification of P/M preforms.

Forgeability

Forgeability is a term that has a variety of definitions. The definition that will be used here is the ability of the metal or alloy to deform without failure, re-

TABLE I

Measured Density and Calculated Densities from True Strain Measurements
for Forged P/M Preforms

Material	Preform Density	Range of ϵ_x		Range of % $\Delta\rho$		Average % $\Delta\rho$
		From	To	From	To	
1020	5.3	0.08	0.79	0.7	6.6	3.5
	6.2	0.07	0.94	−1.2	2.9	0.8
	7.2	0.06	0.81	−0.6	1.6	0.3
1030	5.3	0.08	0.79	0.3	5.4	3.1
	6.2	0.07	0.90	0.5	4.3	2.6
	7.2	0.06	0.76	−3.1	−0.7	−1.6
1040	5.3	0.08	0.76	0.2	5.4	1.7
	6.2	0.07	0.76	−1.3	2.6	0.1
	7.2	0.06	0.68	−2.4	0.0	−1.5
4620	5.3	0.08	0.96	−2.0	3.5	0.7
	6.2	0.06	0.80	−2.5	3.4	0.1
	7.2	0.05	0.73	−1.6	0.3	−0.7
4630	5.3	0.07	0.59	−0.5	4.6	2.2
	6.2	0.06	0.76	−2.0	−0.3	−1.1
	7.2	0.06	0.72	−1.4	1.5	0.3
8620	5.3	0.08	0.61	0.3	3.4	1.9
	6.2	0.07	0.86	−3.9	−0.7	1.8
	7.2	0.05	0.78	−3.4	−0.7	1.9

gardless of the forging pressure requirements. A variety of tests have been used
to assess this property for conventional cast or wrought materials. These tests
include:

1. Hot Twist test
2. Cylinder Upset test
3. Notched Bar Upset test
4. Hot Impact test
5. Tension tests
6. Compression tests

No systematic study of the forgeability of P/M preforms has been reported. As
listed above, the results of compression tests are an indication of forgeability.
In the deformation-densification work that has been reported, uniaxial compression and plane-strain compression tests were made [6,15]. These results indicated that it was possible to deform P/M preforms a significant amount without
cracking. Typical examples of the large amounts of deformation possible are
shown in Figure 18 for a 1018-type steel. True compressive strains of 0.7–1.0
are common. However, torsion tests usually correlate better with maximum forg-

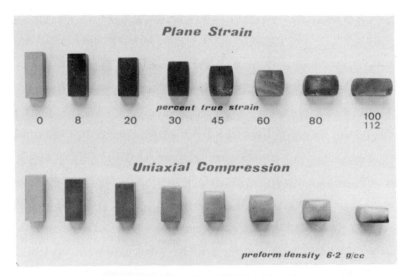

Figure 18. Deformation of 1018 steel preforms.

ing temperature than some of the other types of tests. Therefore, a program was initiated to determine the torsion characteristics of P/M preforms [16]. Tests are being conducted on square and circular cross-section specimens. A square cross-section specimen is shown in Figure 19(a), with three fine-wire thermocouples spot-welded in place. The thermocouples provide an accurate indication of specimen temperature at the time of test; they also are the sensing couples for the controllers for the three-zone quartz-lamp furnace. The entire testing assembly is shown in Figure 19(b), with a close-up of the specimen in the furnace. With this apparatus, it is possible to heat the specimen to 2000°F in less than one minute. A high-speed optical recorder is used to measure the torque and amount of twisting. Some of the preliminary results of this program are shown in Figure 20. These data indicate that forgeability increases with increasing preform density. It also indicates a trend of increasing forgeability with increasing temperature, with the maximum forging temperature decreasing with decreasing density. For example, a fully dense wrought 1040 material exhibits a maximum at about 2250°F [17]. The 7.2 g/cm^3 density preforms may be peaking out at about 2000°F. The 6.7 g/cm^3 density preform peaks out at 1900°F.

The 7.2 g/cm^3 and 6.7 g/cm^3 preforms exhibit a dip in their curves at 1500–1600°F. This type of behavior is observed for conventional wrought materials as well, and is related to the α-to-γ transformation that occurs in this temperature range.

Plots of maximum torque versus temperature are shown in Figure 21. A general trend of decreasing torque with increasing forging temperature is observed.

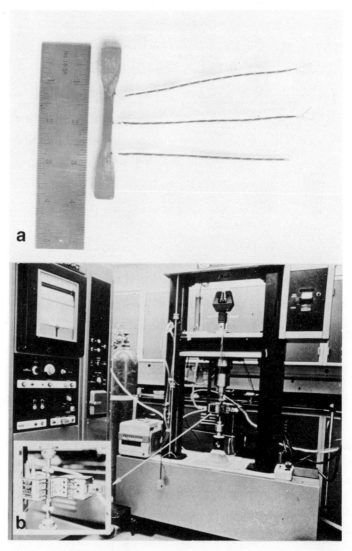

Figure 19. Torsion testing assembly and specimen. (a) Specimen with thermocouples attached; (b) testing assembly.

However, at about 1800°F there is a leveling off of the curves for the 7.2 and 6.7 preforms. Since forgeability of the 7.2 preforms increases only slightly above this temperature (Figure 20), there may be no advantage in forging above 1800°F for this material at this preform density.

The data shown in Figures 20 and 21 are for specimens protected from oxida-

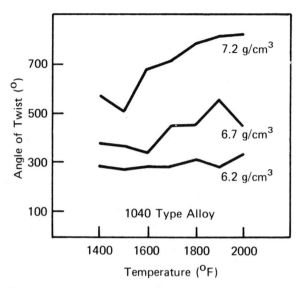

Figure 20. Hot torsion test, angle of twist vs. temperature (specimens tested in argon atmosphere).

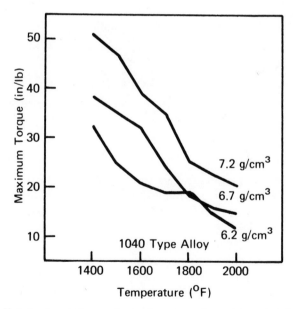

Figure 21. Hot torsion tests, maximum torque vs. temperature (specimens tested in argon atmosphere).

tion by an argon atmosphere in the furnace. In order to determine the effect of the degree of internal oxidation on forgeability, tests were made where heating was accomplished in an argon atmosphere; the argon then turned off and air introduced for a fixed time prior to testing. Some preliminary results of the tests made at 1800°F are shown in Figure 22. It can be seen that forgeability de-

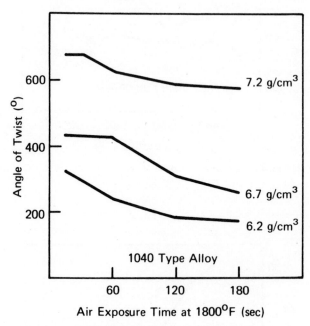

Figure 22. Hot torsion tests, angle of twist vs. air exposure time at 1800°F (specimens heated in argon, exposed to air before testing).

creases with increasing exposure time (internal oxidation). It is interesting to note that for the 7.2 g/cm³ preform, forgeability decreases after an exposure of about 30 seconds. This correlates well with the tolerable exposure time of 33 seconds shown in Figure 14.

It has been shown that forgeability (angle of twist) is affected by temperature, preform density, and degree of oxidation. Another factor that should affect at least the cold forgeability is the degree of sintering. Torsion data for atomized iron preforms sintered at 1700°F and 2050°F are presented in Table II. These data indicate lower forgeability with lower sintering temperature.

TABLE II

Room-Temperature Torsion Tests for 1700°F and 2050°F
Sintered Atomized Iron Preforms

Sinter (30 min)	Sinter Density (g/cm^3)	Max. Torque (in.-lbs)	Angle of Twist at Failure ($^\circ$)
1700°F	6.7	87.0	241
	7.2	109.5	337
2050°F	6.7	89.0	310
	7.2	114.5	410

Properties of P/M Forgings

Properties of Cold Forgings

1. *Impact Strength*. The effect of porosity and inclusion content on mechanical properties will depend on the amount or porosity and inclusions present and also on the particular property being measured. Dynamic properties, such as impact and fatigue strengths, would be expected to be more sensitive to porosity than static properties, such as tensile strength, transverse rupture strength, hardness, etc. For example, the effect of porosity (density) on the impact strength of cold forged atomized iron preforms is shown in Figure 23. At densities only slightly lower than 7.8 g/cm^3 there are significant decreases in impact strength. The specimens used for this work were cut from cold forgings made by plane-strain deformation of $2'' \times 2'' \times 5''$ preforms (6.2 g/cm^3) to the densities shown by the data points.

In order to illustrate the importance of flow on developing maximum properties, $2'' \times 2'' \times 5''$ preforms were made to a 7.2 g/cm^3 density and a 5.7 g/cm^3 density [18]. Both preforms were forged to about 7.78 g/cm^3. The details of the deformations are shown in Figure 24. Four Charpy V-notch impact specimens were machined from each forging, as indicated by Figure 25. The results of the impact test with the fractured specimens are shown in Figure 26. These data indicate that the forging made from the 5.7 g/cm^3 preform (which had to have a greater amount of deformation) had uniformly high impact properties. However, the forging made from the 7.2 g/cm^3 preform (which had a lesser amount of over-all deformation) had a variation in properties with respect to specimen location. The 1 and 4 (top and bottom) specimens, taken from the outer portion of the forging where the amount of flow was greater, had properties equivalent to those of the forging made from the lower density preform. However, the 2 and 3 specimens, taken from the inner portion of the forging (made from the 7.2

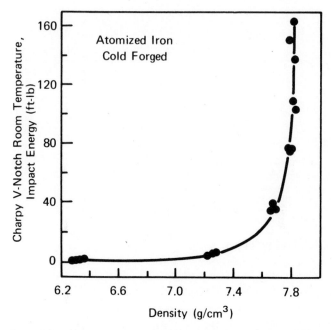

Figure 23. Impact properties of cold forged atomized iron preforms.

g/cm³ preform), had much lower properties. The specimens that had the lower impact properties (2 and 3) experienced less flow than the outer specimens (1 and 4) because of the extent of the dead-metal zone in the center portion of the forging. Thus, it is important to achieve a high density but it is equally impor-

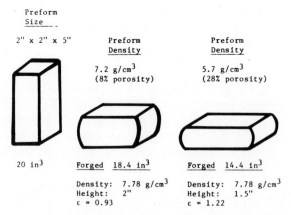

Figure 24. Preform forging relationship for cold forged 7.2 g/cm³ and 5.7 g/cm³ density preforms.

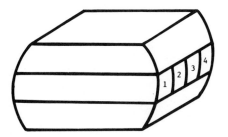

Figure 25. Specimen location in powder preform forgings.

tant to have some minimum amount of flow if maximum properties are to be achieved. Similar results have been found in hot forging.

In order to evaluate the effect of inclusions on the impact properties, various amounts of natural inclusions were blended into atomized alloy powder. The mixture was pressed, sintered, cold forged to full density, and then resintered. The results of V-notch impact tests on these specimens are shown in Figure 27. These data indicate that increasing the inclusion content, even in small quantities, seriously affects the impact properties.

2. *Hardness.* In contrast to the abrupt change in impact properties with density, the hardness increases gradually as the density is increased. The variation in hardness with density for preforms deformed in plane strain are shown in Figure 28 for atomized iron, 1020, 4620, and 4630 materials.

3. *Tensile Properties.* In order to determine the tensile properties of cold forged preforms, a modified 4640 material (1.8 Ni–0.5 Mo) was pressed and sintered to 7.2 g/cm^3. Preforms were cold forged to full density and then re-

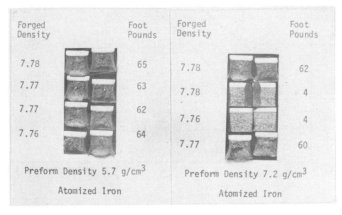

Figure 26. Effect of flow on impact energy and fractures of forged atomized iron preforms.

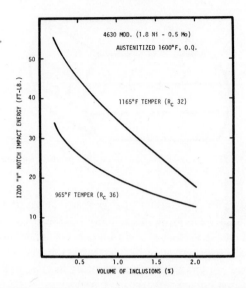

Figure 27. Effect of inclusion on impact properties of 4630 P/M forgings.

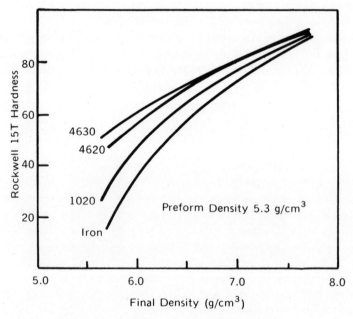

Figure 28. Hardness vs. density for P/M forgings.

heated to 2050°F for 30 minutes. Tensile properties after reheating and quenching and tempering are shown in Table III.

TABLE III

Tensile Properties of Cold Forged Modified 4640 Material

Condition	UTS (ksi)	% R.A.
Reheated	104.6	36.1
1550°F Oil Quench,	194.3	35.8
800°F, 1 hr		

These tensile properties indicate that high strength with good ductility can be achieved with cold forged preforms.

In summary, iron, iron-carbon, and low-alloy steels may be cold formed to full density. At full density, tensile strength, ductility, and impact strength are good if there is adequate flow in forging.

Properties of Hot Forgings

As in the case of cold forged P/M preforms, residual porosity and inclusions will have a degrading effect on properties of the forgings. The magnitude of this effect will depend on the particular property being measured. In general, P/M preforms forged to full density exhibit tensile properties at least as good as conventional wrought materials. Impact properties have less variation with specimen orientation than conventional wrought materials. However, more work is needed in this area. Fatigue properties as measured on polished test specimens indicate that P/M forgings have lower fatigue limits. However, tests conducted on actual forged components indicate that P/M forgings have comparable fatigue limits [20]. There can be very significant differences in surface finish between conventional and P/M forgings [21]. The smoother finish of P/M forgings could well account for the more favorable comparison in fatigue properties when actual components are tested.

1. *Tensile Properties.* Tensile properties and photomicrographs of P/M forged 1040, 4140, and 4340 steels are shown in Figure 29. The properties of the P/M forgings are shown in the numerator of the fraction, with average properties of conventional equivalent wrought materials shown in the denominator. The photomicrographs indicate that the P/M forgings are relatively clean. The properties of the P/M forgings compare favorably with conventional wrought material when heat-treated to the same hardness level. However, if maximum density is not achieved, the ductility may be significantly reduced. In Figure 30, the

202 H. W. ANTES

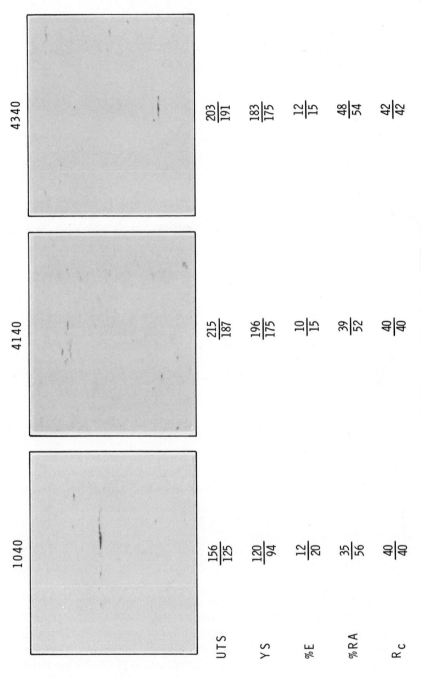

Figure 29. Structure and properties of P/F steels.

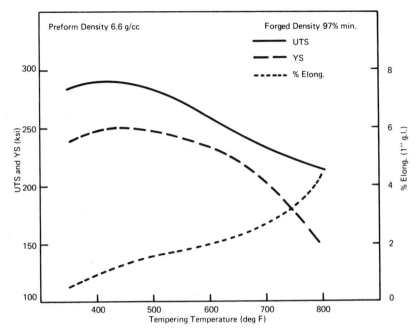

Figure 30. Powder-forged 4665 [22].

tensile properties of a 4665-type material are plotted as a function of tempering temperature [22]. The ductilities shown in Figure 30 are somewhat lower than those in Figure 29. Some of this difference may be because the data in Figure 30 are reported for forged density of 97 percent min. and full density may not have been achieved.

Various investigators have reported the fact that good tensile properties can be achieved in P/M forgings if full density is achieved [7,20,22-25].

It has been reported that a linear correlation exists between hardness and ultimate tensile strength in much the same manner as conventional wrought materials [23]. These data are shown in Figure 31, and indicated that

$$\text{UTS (ksi)} = \frac{\text{Vicker hardness}}{2.1}$$

Most of the tensile properties are reported for confined-die or closed-die forging. However, P/M preforms forge quite well in open-die forging. As an example of the high forgeability, $2'' \times 2'' \times 5''$ preforms ($\rho = 6.2 \text{ g/cm}^3$) were hammer-forged at 1900°F. Billets were either sealed in low-carbon steel cans (electron-beam welded in cans) or coated with a water suspension of graphite, heated in air to 1900°F, and hammer-forged to approximately $1''$-square bar stock. Actual forgings with a preform are shown in Figure 32. Tensile properties of these

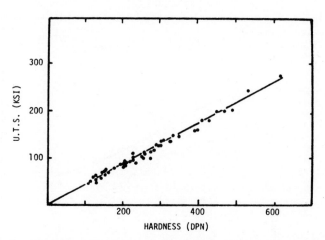

Figure 31. Correlation of hardness with ultimate tensile strength for P/M forgings.

forgings (8627 material) are shown in Table IV. It was reported that 8630 pre-forms hot repressed to 98 percent of theoretical density have tensile strengths equivalent to conventional wrought material, but the elongation was somewhat lower [26]. The impact and fatigue properties of this material were also good. These properties will be discussed later.

2. *Impact Strength.* Impact properties of hot forged atomized iron proper-ties are shown in Figure 33 [27]. Although the data in Figure 33 are for U-

Figure 32. Open-die forged 8627-type material. Top, $-2'' \times 2'' \times 5''$ preform; center, forging from graphite-coated preform; bottom, forging from canned preform.

TABLE IV

Tensile Properties of Open-Die Hammer-Forging 8627 P/M Preforms

Material	UTS (ksi)	YS (ksi) (0.2% off)	% E (2" g.l.)	% R.A.
Canned 8627	107.5	86.5	17.0	44.0
Coated 8627	110.0	88.7	16.0	39.7

notched impact specimens and those for cold forged atomized iron were for V-notched specimens (Figure 23), identical behavior was observed.

It was mentioned previously that the type and amount of deformation may have a significant influence on the dynamic properties of P/M forgings. Some data have been reported in the literature that indicate low impact properties for preforms supposedly forged to full density. If the preforms have essentially zero porosity and impact properties are low, it may be the result of high inclusion content or inadequate flow during deformation. The term "flow" is not restricted to mean external lateral flow, since low-density preforms provide significant amounts of internal freedom for flow. Relatively good impact properties have been reported for hot repressed 8630 preforms with approximately 1–2 percent residual porosity [26]. Some of these data are shown in Table V.

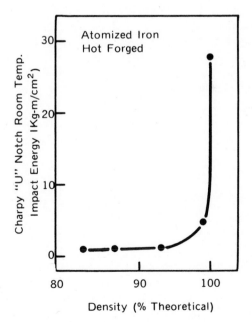

Figure 33. Impact properties of hot forged atomized iron preforms [25].

TABLE V

Properties of Hot Repressed 8630

YS (ksi)	UTS (ksi)	% E (1 in.)	% R.A.	Mod. of Elast. (x10^6)	R.T. Charpy V-Notch (ft-lb)
174.0	233	7.0	18.0	26.0	8.5

Wrought 8630 material at the same hardness would have an impact value of approximately 16 ft-lb. The impact properties of the P/M forgings (full density) reported in Figure 29 are given in Table VI.

TABLE VI

Impact Properties of P/M Forged Steel

Material	UTS (ksi)	R.T. Charpy V-Notch (ft-lb)
1040	156	8
4140	215	12.5
4340	203	26.5

The 8630 material [26] not at full density has relatively good impact strength. The impact strength of the alloys shown in Figure 29 and Table IV would probably increase if the carbon content was decreased to 0.3, as in the 8630 material. The 8627-type material shown in Figure 32 has the impact properties shown in Table VII.

·TABLE VII

Impact Properties of Open-Die Hammer-Forging 8627 Preforms

Material	UTS (ksi)	Charpy V-Notch (ft-lb)
Canned 8627	107.5	24
Coated 8627	110.0	15

More complete data are needed on the effects of processing variables on impact strength and impact transition behavior of P/M forgings.

3. *Fatigue Strength.* A limited amount of fatigue property data for P/M forgings has been published. Fatigue tests made on the actual P/M forged components compare favorably with conventional forged components, probably

because of the better surface finish of the P/M forgings. However, when polished fatigue-test specimens are used for both powder-forged and conventionally forged material, the P/M forged material does not compare as favorably with the conventionally forged material. More work is also needed in the area of the effect of processing variables on the fatigue properties of P/M forgings.

4. *Hardenability.* A recent study has indicated that existing production sintering facilities are adequate for alloys containing Cu, Ni, Mo, and Cr. Manganese may be used if the concentration is kept under 1 percent [28]. Grain size may have an adverse affect on hardenability if it is too fine. Jominy hardenability curves for several P/M forged steels (ASTM GS7-8) are shown with H-bands for AISI steels in Figure 34.

Summary

Three basic types of P/M forging processes have evolved. One process is best described as a hot repressing process with no lateral flow. The second is the flashless process with lateral flow and is carried out in confined or trap dies. The third is similar to conventional closed-die forging with limited flash produced.

Some aspects of inclusion and void behavior during deformation were covered. The importance of both shearing and hydrostatic stresses during deformation was presented. These stresses affect residual porosity and void formation associated with inclusions. Preform design and fabrication is one of the most critical steps of the P/M forging process. The preforms must be protected in the case of hot forging so that important alloying elements are not preferentially oxidized or degrading oxides are not entrapped in the forging. There is some natural protection offered to the preform by the reducing atmosphere used in the heating process. However, the tolerable exposure time in air before forging is sensitive to preform density and to forging temperature. Precoating of preforms with graphitic materials have been used to extend the tolerable exposure time.

The type, as well as the amount, of deformation affects the rate of densification and the properties of P/M forgings. Temperature and the related effect of speed of deformation reflect on the ease of deformation. Lubrication is important in that it affects die life, surface densification, forging load, and homogeneity of deformation.

Some preliminary results of torsion tests were presented to give some insight on the problem of forgeability.

Mechanical properties of cold forged preforms are good at full density, and compare favorably with conventional wrought materials. The properties are sensitive to residual porosity, especially dynamic properties such as notched impact strength. The properties of hot forged P/M preforms compare favorably

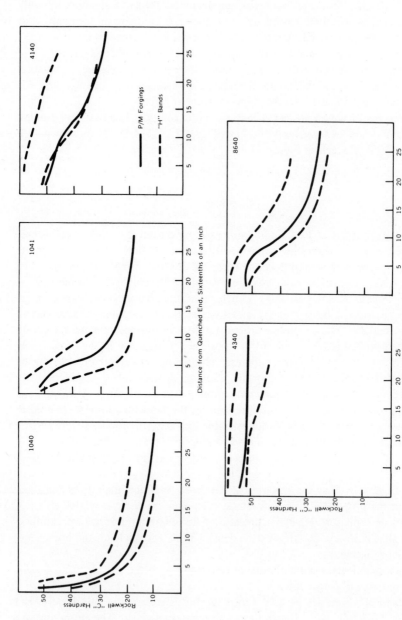

Figure 34. Jominy end-quench hardenability curves for powder-forged steels.

with conventional forgings. Results were presented that indicate that P/M preforms exhibit good forgeability in open-die hammer-forging.

Hardenability of P/M forgings will fall within the H-bands, provided that alloying elements are not oxidized and that grain size is not too fine.

References

1. Halter, R.F., "Pilot Production System for Hot Forging P/M Preforms," International Powder Metallurgy Conference, New York, July 1970, in *Modern Developments in Powder Metallurgy*, Vol. 4, *Processes*, H.H. Hausner, ed., New York: Plenum Press (1971), 385.
2. Takeya, Y., Hayasaka, T. and Kamata, "Effect of Various Types of Iron Powder on Properties of High Density Fe-Based Alloys," Brutcher Translation, no. 7940, Altadena, Calif.: Henry Brutcher Technical Translations.
3. Reprinted with permission of the Wakefield Bearing Company.
4. Reprinted with permission of Engineered Sinterings and Plastics, Incorporated.
5. Reprinted with permission of Great Lakes Forging, Limited.
6. Antes, H.W., "Cold Forging Iron and Steel Powder Preforms," International Powder Metallurgy Conference, New York, July 1970, in *Modern Developments in Powder Metallurgy*, Vol. 4, *Processes*, H.H. Hausner, ed., New York: Plenum Press (1971), 415.
7. Brown, G.T. and Jones, P.K., "Experimental Aspects of the Powder Forging Process," International Powder Metallurgy Conference, New York, July 1970, in *Modern Developments in Powder Metallurgy*, Vol. 4, *Processes*, H.H. Hausner, ed., New York: Plenum Press (1971), 369.
8. Huseby, R.A., "Forgings from P/M Preforms," ASM Powder Metallurgy Conference, Chicago, February 1970.
9. Reprinted with permission of the Heppenstall Company.
10. Vernia, Peter, "Short Cycle Sintering by Induction," International Powder Metallurgy Conference, New York, July 1970, in *Modern Developments in Powder Metallurgy*, Vol. 4, *Processes*, H.H. Hausner, ed., New York: Plenum Press (1971), 475.
11. Cook, J.P., "Degradation of P/M Forging Preforms During Interim Exposures in Air," Metals Engineering Congress, AMPI–ASM Powder Metallurgy Conference, Cleveland, Ohio, October 1970, in *Fall Powder Metallurgy Conference Proceedings*, New York: MPIF (1970), 237.
12. Dieter, G.E. Jr., *Mechanical Metallurgy*, New York: McGraw-Hill Book Company, Inc. (1961), 60.
13. Kuhn, H.A. and Downey, C.L., "Deformation Characteristics and Plasticity Theory of Sintered Powdered Materials," *Int. J. Powder Met.*, 7 (1971), 15.
14. Bockstiegel, G., "A Study of the Work of Compaction in Powder Pressing," *Powder Met. Int.*, 3 (1971), 17.
15. Antes, H.W., "Cold and Hot Forging P/M Preforms," SME Technical Report EMR 71–01, 1971.
16. Stockel, P., "High Temperature Torsion Tests of Atomized Powder Preforms," Hoeganaes Corporation Technical Report no. 1–72, 6 January 1972.
17. "Evaluating Forgeability of Steels," published by the Timken Roller Bearing Company, Canton, Ohio.

18. Mayer, K.H., Research in progress, to be published by the Hoeganaes Corporation, Riverton, New Jersey.
19. Muzik, J., Research in progress, to be published by the Hoeganaes Corporation, Riverton, New Jersey.
20. Cull, G.W., "Mechanical and Metallurgical Properties of Powder Forgings," *Powder Met.* 13 (1970), 156.
21. "P/M Forgings," Hoeganaes Bulletin, Form 210, Hoeganaes Corporation, February 1971.
22. Knopp, W.V., "Properties of Hot Forged P/M Steels," *Fall Powder Metallurgy Conference Proceedings*, New York: MPIF (1969), 9.
23. Cundill, R.T., Marsh, E. and Ridal, K.A., "Mechanical Properties of Sinterforged Low-Alloy Steels," *Powder Met.*, 13 (1970), 165.
24. McGee, S.W., "Design in Powder Metallurgy and Sinterforging," SME, Mississippi Valley Section, Winter Meeting, Waterloo, Iowa, 21 January 1971.
25. Bargainnier, R.B. and Hirschhorn, J.S., "Forging Studies of a Ni–Mo P/M Steel," *Fall Powder Metallurgy Conference Proceedings*, New York: MPIF (1970), 191.
26. Pietrocini, T.W., "Hot Densification of P/M Alloy 8630," *Precis. Metal*, 29 (1971), 58.
27. Hirschhorn, J.S. and Bargainnier, R.B., "The Forging of Powder Metallurgy Preforms," *J. Metals*, 33 (1970), 21.
28. Eloff, P.C. and Kaufman, S.M., "Hardenability Considerations in the Sintering of Low Alloy Powder Preforms," *Powder Met. Int.*, 3 (1971), 71.

11. Hot Isostatic Pressing of High-Performance Materials

H. D. HANES
Battelle, Columbus Laboratories
Columbus, Ohio

ABSTRACT

Hot isostatic pressing (HIP) is rapidly achieving a place as a production process for a variety of high-performance powder-metallurgy materials. A general description of the process is given. Examples of powder-metallurgy shapes made by HIP are discussed. Properties of various high-performance materials, including beryllium, titanium, tool steels, and carbides, are shown. Current and future applications of the process are also presented.

Introduction

Hot isostatic pressing (HIP) has grown over the past 15 years from a laboratory curiosity to a viable process capable of producing large-scale, high-quality powder-metallurgy products on a production basis. Many investigators at various facilities have made significant contributions to this rapid growth. Certainly, it is beyond the ability of any single author to review adequately all of these contributions in one chapter. High-pressure equipment has progressed from small, laboratory-sized pressure vessels with long turn-around cycles to large-diameter equipment capable of handling large charges in short order. Designers of furnaces for this high-pressure equipment have constructed heaters that occupy a large percentage of the pressure-vessel volume, thereby insuring a maximum output for a given size of pressure vessel. Unique methods have been devised for HIP container fabrication which enable pressing of components to nearly net configuration in relatively low-cost, reliable containers. Utilizing these methods, structures have been fabricated from several materials with properties equivalent to, and sometimes superior to, those produced by other methods. Lest this sound like the powder metallurgist's panacea, let me hasten to add that there are many materials and/or structures that cannot be fabricated by this technique because of economic or technical factors, at least within the current state of the art.

211

The objectives of this chapter will be fourfold. First, a generalized description of the HIP process will be given. Second, examples of unique structures fabricated by this method will be cited. Third, properties of high-performance engineering materials fabricated by this technique will be discussed. Finally, current and future potential of HIP as a production process will be discussed.

Although this paper deals exclusively with P/M applications of hot isostatic processing, it should be noted that this type of equipment has been utilized for other methods of materials processing. These techniques include isostatic diffusion bonding of wrought components into structures (termed gas-pressure bonding), healing of internal defects in castings, and forming of sheet components under hot isostatic pressure.

Many of the generalized arguments presented for the utilization of a P/M approach also apply to HIP. The most potent of these is the ability to control microstructure, chemistry, and/or crystallographic texture in a final product; thus, the potential of controlling final properties exists through P/M processing. HIP actually extends this potential because consolidation can be normally achieved at a lower temperature than more conventional processing. Because of this, more latitude exists in controlling microstructure during processing. The sealed HIP container allows the addition or retention of volatile constituents during processing, so chemistry can be accurately controlled.

Additionally, HIP offers the potential of fabricating geometries that are difficult to achieve by more conventional processing, such as components with divergent angles, irregular internal or external contours, thin walls, long aspect ratios, etc. Essentially full density (or controlled levels of porosity, if desired) can be achieved in most materials. Radioactive or toxic materials can be readily accommodated within the sealed container. Also, as will be discussed in this chapter, full-wrought properties can be achieved in the as-HIP condition in some materials.

Description of the HIP Process

Hot isostatic pressing is the consolidation of particulate materials by gas (or fluid) pressure at an elevated temperature. The components to be consolidated are hermetically sealed in a container which can be deformed at the elevated temperature, thus compacting the contained powders. The powders are preformed by a variety of techniques including die-pressing, cold isostatic pressing, or direct vibratory packing into the container. Prior to final sealing of the container, the powders are normally evacuated at room or elevated temperature or pretreated as required. The canned components are then loaded into the pressure vessel, processed to predetermined parameters of time, temperature, and pressure, and removed from the vessel. Depending on cycle parameters or equipment limitations, components can be preheated or removed hot from the cycle for post-pressing thermal treatment. Figure 1 schematically illustrates the basic steps in assembling and processing a component for HIP.

Shown in Figure 2 are the basic components of a high-pressure system utilized for HIP. These components include the gas-storage container and compressor (pressurization system), the control panel and furnace (thermal system), and the pressure vessel in which the compaction is done. Various penetrations are made through one or both of the closures and/or sidewalls for pressurization lines, furnace power leads, instrumentation, and supplemental cooling of the pressure-vessel body and closure seals.

The primary limiting factors on the rate of production arise from the capacity of the pressurization system and the closure mechanism of the pressure vessel. The limitations of the thermal system are probably of secondary importance to the rate of production, since components can be preheated external to the pressure vessel if the production scheme dictates. Pressurization is normally achieved through the use of a multiple-stage compressor. Higher rates can be achieved through the use of pressure accumulators or by filling directly from storage. Helium and argon are the usual pressurizing gasses, but nitrogen and air

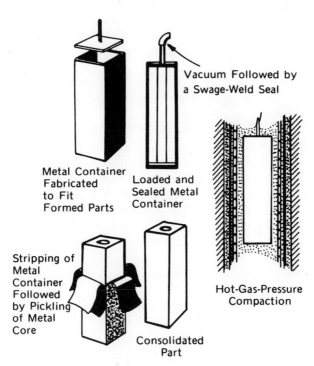

Vacuum Followed by
a Swage-Weld Seal

Metal Container
Fabricated
to Fit
Formed Parts

Loaded and
Sealed Metal
Container

Stripping of
Metal
Container
Followed
by Pickling
of Metal
Core

Hot-Gas-Pressure
Compaction

Consolidated
Part

Figure 1. Schematic diagram of hot-isostatic pressing process.

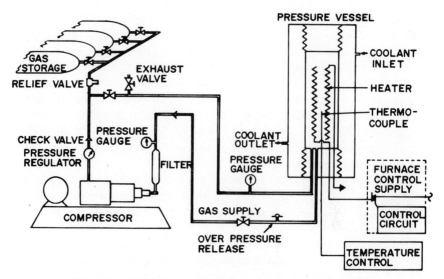

Figure 2. Simplified schematic for hot-isostatic pressing system.

have also been used. Rapid access to the pressure vessel is also critical to achieving fast turn-around times in production. Different manufacturers have used various methods for achieving rapid closure.

Typical Shapes Fabricated by HIP

Following the general process scheme outlined above, many different shapes of varying degrees of complexity have been fabricated. It should be noted that many of these components are experimental items that have been generated for specialized applications. Nevertheless, they demonstrate the degree of sophistication that can be achieved by the HIP process. This portion of the chapter will be restricted to relatively complex geometries, since simple geometries are within the state of the art.

HIP Shapes with Complex External Detail

The ability to produce nearly net shapes by HIP with complex external geometry has been demonstrated. Figure 3 shows a beryllium gyro air-bearing sleeve in the pressed, partially machined, and fully machined conditions [1]. A visual comparison of these three steps shows the reduction of machining and expensive materials loss. A more vivid demonstration of the complexity of

Figure 3. Beryllium air-bearing sleeve in as-HIP, rough-machined, and final-machined states.

shapes that can be achieved is shown in Figure 4. The beryllium inner gimbal, measuring roughly 10 inches on a side, represented a materials savings of approximately 65 percent [2]. The hot isostatically pressed drill bit shown in Figure 5 is a shape of more general interest. Although this component was made from a nickel-base alloy, one can envision the same approach being used to generate a similar semifinished shape from a carbide or a high-alloy steel.

To generate these types of shapes, the powder is first preformed by cold isostatic pressing. Then, the preform is loaded into a HIP container with a regular geometry and the space surrounding the preform is filled with a pressure-transmitting compound. In the case of the beryllium components, sodium chloride was used. The pressure transmitter is chosen on the basis of its chemical compatibility with the container and part, relatively low cost, and ability to transmit the pressure evenly from the container to the part being consolidated.

HIP Shapes with Complex Internal Geometries

A variety of shapes have been fabricated by HIP, whose internal cavities are formed by mandrels that are removed either chemically or mechanically. The mandrels are selected on the basis of metallurgical compatibility with the part

Figure 4. Beryllium inner gimbal HIP preform in early stages of machining.

(or a suitable barrier is used), a reasonable match of thermal-expansion coefficients, and dimensional stability of the mandrels at the HIP temperature. In several instances, extremely close (1-2 mil) tolerances have been achieved in these components.

Figure 6 shows a tungsten grid fabricated for a specialized application. Molybdenum mandrels were accurately held at each end by photoetched grids. The spherical tungsten particles were vibratorily filled into the cavity defined by the container and the mandrels. After HIP, the outside surface was machined to

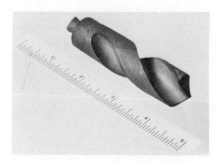

Figure 5. HIP preform of a drill bit.

Figure 6. Tungsten grid hot-isostatically pressed on leachable molybdenum mandrels.

final configuration and the mandrels were selectively leached from the structure. Dimensional analysis revealed that the channels and webs were within 2 mils of the design size. A similar approach [3] was utilized to fabricate the prototype nickel-alloy injector plate shown in Figure 7.

Figure 8 illustrates beryllium seamless pressure-storage bottles fabricated by

Figure 7. Nickel-alloy injector disk HIP with internal coolant passages.

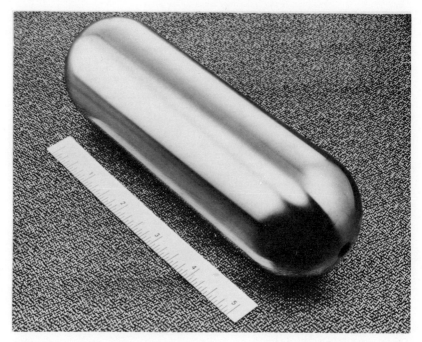

Figure 8. Beryllium seamless pressure vessel hot isostatically pressed around a leachable mandrel.

pressing powder around a mandrel which was subsequently leached from the internal cavity through the fill port. Similar structures fabricated from titanium–6aluminum–4vanadium (Ti–6–4) powders in the solutionized and aged condition were pressurized to 154,000 psi maximum fiber stress without failure.

Thin-Walled Beryllium Structures by HIP

Large, hollow beryllium structures have been produced on a pilot-production scale by HIP [4]. These structures were approximately 16.5 inches in diameter at the base and measured over 42 inches in length. The wall thickness varied from 0.5 to 1.2 inches. On a program originally sponsored by Lockheed Missiles and Space Company and later by Kawecki-Berylco, these components were pressed to nearly net ID configuration on permanent mandrels. By proper selection of a suitable barrier layer, it was possible mechanically to extract the mandrel for re-use.

Using the same type of technology, the HIP process is currently being developed on an Air Force program for fabricating thin-walled beryllium conical frustra [5,6]. To date, subscale hardware (8-inch-diameter base by 18-inch length) has been fabricated on this program (see Figure 9). Structures of this type with wall thicknesses down to 0.060 inch have been pressed and removed from the HIP mandrel with no cracking. It is estimated that structures of this type can be pressed to 80–85 percent of net configuration.

Properties of High-Performance Materials

Beryllium

Much research has been conducted at Battelle on the application of HIP to the fabrication of various beryllium components, as has been previously noted. This is a logical occurrence, since most of the forms of beryllium used for struc-

Figure 9. Thin-walled (0.060″) beryllium cone after machining the OD.

tural applications are from P/M sources, and the potential economic payoff justified the use of HIP in its earlier stages of development. In the case of gyro applications, the fact that HIP produced fully dense material that would accept high surface finishes also justified the approach.

Most recently, the development of the HIP process has been aimed at fabricating structural components to nearly net configuration. As a result of these programs sponsored by the Navy (through LMSC) and the Air Force and by industrial concerns, considerable insight has been gained into the powder-metallurgical behavior of beryllium. Table I lists several of the grades of beryllium powder that have been consolidated by this technique and typical properties achieved. It should be noted that in some of the grades, insufficient work was done to optimize the processing parameters. As a result, some of the elongation figures appear somewhat low.

Generally speaking, the properties listed in Table I are significantly higher than those of comparable grades of vacuum-hot pressed block. This is attributed to the fact that the HIP process is accomplished at temperatures as much as $300°F$ lower than conventionally processed material, resulting in a finer grain size (see Figures 10 and 11). It has been reported by Lidman and Younger [7]

TABLE 1

Typical Properties of Beryllium Processed by Hot Isostatic Pressing

Type Material	HIP Parameters	Typical Tensile Properties			Density, G/cc
		Y.S. (ksi)	UTS (ksi)	Elong. (%)	
KBI[a] P-8	3 hrs at 1675 F, 10,000 psi	35.6	54.9	2.53	1.856
G.A.[b] GB-1	3 hrs at 1675 F, 10,000 psi	38.3	54.4	1.90	1.861
G.A. GB-2	3 hrs at 1700 F, 10,000 psi	49.0	61.0	1.66	1.863
BB[c] S100	3 hrs at 1400 F, 10,000 psi	48.1	59.7	1.21	1.849
BB S100	3 hrs at 1675 F, 10,000 psi	48.1	57.6	1.17	1.855
BB S200	3 hrs at 1725 F, 10,000 psi	69.4	76.6	0.76	1.862
KBI–HPEFP[d]	3 hrs at 1675 F, 15,000 psi	37.0	58.0	4.25	1.854
Specimens from Hollow Structures					
G.A. GB-1 ⊥to press direction	3 hrs at 1675 F, 10,000 psi	41.6	56.3	3.4	1.861
G.A. GB-1 ⊥ (miniature specimen)	3 hrs at 1675 F, 10,000 psi	37.6	53.2	2.1	1.861
‖ (miniature specimen)		38.5	51.1	1.7	1.861

[a]Kawechi-Berylco Industries, Inc.
[b]General Astrometals (now Yonkers Div. of KBI).
[c]Brush Beryllium Corporation.
[d]HPEFP = high-purity electrolytic flake powder.

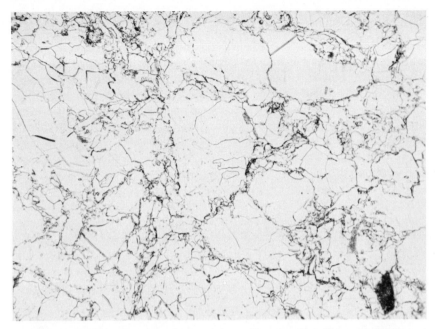

Figure 10. Typical microstructure of a commercial grade of beryllium after hot-isostatic pressing. 500X.

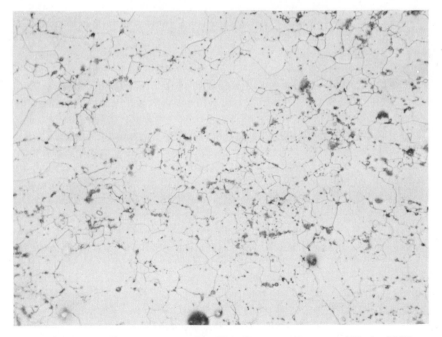

Figure 11. Microstructure of a commercial grade vacuum–hot pressed block. 500X.

that the properties in large structures produced by HIP are more consistent from part to part and between various points within these parts than for hot pressed block. Mueller [6] has confirmed that this consistency of properties extends to shells that are pressed to wall thicknesses of 0.060 inch. As would be expected, there was a small degree of mechanical anisotropy relative to the direction of pressing in a hollow shell pressed on a rigid mandrel. Mechanical tests taken in the thickness direction (roughly analogous to the pressing direction in vacuum–hot pressed block) showed that the elongation was well above the minimum specification for hot pressed block.

Recently, KBI has developed a powder from high-purity electrolytic flake source ingot. This high-purity ingot is swarfed and converted to HPEFP powder by impact attritioning. The resultant powder is probably of the highest purity ever produced on a production basis. It can be noted in Table I that this marked increase in purity has increased the elongation to greater than 4.25 percent with very little decrease in strength level. The development of this new grade of powder, combined with HIP processing, opens a new potential area of improvement of beryllium P/M products.

Titanium–6Aluminum–4Vanadium HIP Processing

In a Battelle-funded program, Harth [8,9] has shown that properties comparable to wrought material can be achieved in HIP Ti–6–4 powder (Table II). Using powder purchased from Nuclear Metals' rotating-electrode process, impurity levels were maintained within ELI specification. The data presented indicate that tensile properties compare favorably with those of wrought material in both as-HIP, annealed, and solutionized and aged (STA) condition. To date, only thermal treatments designed for cast and wrought material have been followed. Optical microscopy, shown in Figure 12, indicates no evidence of interparticle separation in the as-HIP product. The replica electron micrograph shown in Figure 13 reveals a typical ductile fracture. The difference in properties measured at AFML and Battelle on the same lot of material in the STA condition suggests that the HIP product may be very sensitive to cooling rate from the solutionizing condition. The data also indicate that the HIP material can be hot worked and still achieve a good level of properties.

Perhaps the most interesting data, albeit preliminary in nature, are the fracture-toughness values generated by Sajdak and Geisendorfer at AFML [10]. The KlC values generated on both the annealed and STA HIP material are significantly higher than those reported for cast and wrought Ti–6–4. Fatigue values appear to be in the same range as those of the wrought alloy, although the data are too limited to report values at this point.

One other point is worthy of note: there has been no indication of swelling

TABLE II

Tensile Properties on Ti–6Al–4V Consolidated by Hot Isostatic Pressing

Specimen Condition	UTS (ksi)	Y.S. (ksi)	Elong. (%)	R.A. (%)	K1C (ksi √in.)
Annealed					
Typical wrought properties	120–150	120–135	10–20	20–30	55
As-HIP	138–140	130–133	13–19	19–28	
As-HIP + vacuum anneal	135	127	19	37	64.8
Solution Heat-Treated + Aged					
Typical STA wrought properties	155–172	146–162	7–11	25–30	45
As-HIP + STA (900 F)	175–177	162–168	5–8	9–11	
AFML data	156	146	18	29	56.4
Swaged HIP Material					
HIP + 18-percent swage reduction + vacuum anneal	137	130	18	55	
Forged HIP Material					
HIP + 40-percent upset + STA (1100 F)	150	141	22	44	

at elevated temperature due to argon adsorption by the powders as has been previously reported. This was verified by optical and electron microscopy and immersion-density checks on materials which have been solutionized at 1750°F. This was further checked by studying the fusion and heat-affected zones in electron-beam welds in HIP material.

Although the data presented here are preliminary in nature, they indicate that an as-HIP P/M structure could possibly be used in applications heretofore considered only for wrought materials. By most aerospace companies' estimates, approximately 75 percent of the titanium used in making shapes ends up as machining scrap or as parts which are rejected because of internal defects. The HIP approach can potentially decrease this loss significantly. Even if structural engineers would not consider an as-pressed product, the potential exists to make fabrication preforms which will result in more nearly net fabricated shapes.

Tools Steels

Considerable publicity has been given during the past year to the production of a high-grade tool steel by both Crucible Specialty Metals Division of Colt Industries [11] and Stora Kopparberg Corporation [12]. The basic processes

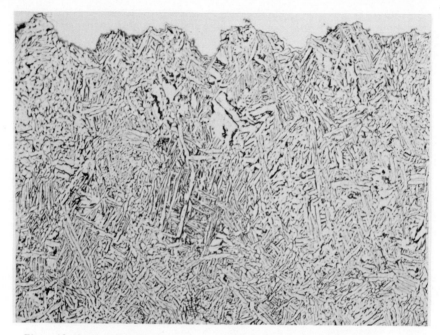

Figure 12. Section through fracture on as-HIP Ti–6Al–4V tensile bar, showing no evidence of interparticle failure. 150X.

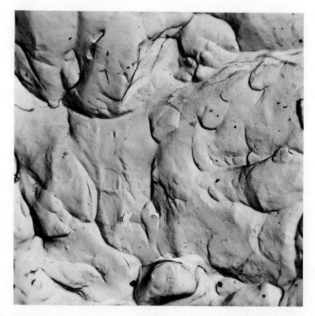

Figure 13. Replica taken for fracture surface on Ti–6Al–4V tensile bar, showing typical ductile fracture. 5100X.

utilized by both companies appear, from the published literature, to be largely the same. The powder, generated by gas atomization to give fine distribution of the primary carbides, is roughly spherical in nature. Because of the morphology of the particles, the product cannot be readily consolidated by more conventional techniques. HIP provides the means to consolidate the gas-atomized particles into billets with a large aspect ratio which can be worked mechanically into machining stock for a wide variety of cutting tools.

Controlling the microstructure of tool steels by P/M processing results in a product that is more responsive to thermal treatments. Both companies have shown that their P/M products are more easily ground to tool shapes and that the resultant tools have significantly longer life. Probably the most significant point to recognize is that the HIP process, which has been heretofore classed as a relatively expensive process, has been successfully applied to the economic production of a relatively low-cost material.

HIP of Carbides

Carbides have been traditionally produced by liquid-phase sintering for a variety of commercial applications such as high-speed cutting tools and precision rolls. This type of processing normally results in a product with residual porosity that can degrade tool life or cause rejection of parts requiring a high surface finish. Hot isostatic pressing is capable of producing fully dense products by either consolidating the green-pressed constituents or removing residual porosity in a presintered piece. Sandvik Steel has indicated that the yield of tungsten carbide rolls for rolling mills can be increased from 5 percent for conventionally processed material to 90 percent for HIP material [13]. Kennametal and Mitsubishi are also planning to produce certain grades of carbide by this method.

Production Potential of HIP

The opening premise of this chapter was that HIP is currently a production process. Several companies, some of which have already been mentioned, are currently producing products by this process. To summarize, the following companies are currently, or soon will be making a product by HIP:

1. Superalloys—Kelsey-Hayes;
2. Tool steels—Stora, Crucible;
3. Carbides—Kennametal, Sandvik, Mitsubishi;
4. Beryllium—Kawecki-Berylco (by end of 1971).

It should be noted that, in the case of tool steels and superalloys, the relative cost of these materials is low and the economics are dependent on the ability to handle a high volume of material.

Certainly, much of the credit for bringing this process to the production stage must go to the designers and producers of HIP equipment. One of the major factors is the design of rapid closures for quick turn-around of products. Figure 14 demonstrates a mechanized interrupted-thread closure on a 5-foot-diameter, 20,000-psi isostatic press manufactured by Midvale-Heppenstall. Similarly, full-threaded closures have been mechanized to allow quick access. Alternative methods to the threaded closure have also been mechanized. One design, by Autoclave Engineers, Incorporated, shown in Figure 15, uses a lateral, tapered plug which is hydraulically driven into position to seal the closure. Pressure vessels designed by ASEA utilize a wire-wound frame to hold the closures in

Figure 14. Interrupted threat closure on 5-foot-diameter, 20,000 psi pressure vessel, (Courtesy of Cohart Industries, Inc.)

Figure 15. Automated pen-type closure mechanism for 25,000 psi operation. (Courtesy Autoclave Engineers, Inc.)

position during the pressure cycle, as shown in Figure 16. All of these designs, which have been shown to be reliable in production service, provide the means for rapid turn-around of a P/M product.

Equally important to the aspect of rapid turn-around is the ability to operate a relatively large-diameter furnace with a well-controlled hot zone and a long operating zone. From the standpoint of capital investment, the relationship of the diameter and length of the furnace to the pressure-vessel dimensions is important. By having the working diameter of the furance as close as possible to the pressure-vessel wall, the capital investment for a given size capability can be minimized. Not only is the size of the pressure vessel affected but the size of supporting pressurization equipment is also minimized. In the Battelle HIP facility, we have routinely operated at temperatures up to 1900°F in a furnace of our own design with a usable hot zone measuring 18 inches ID by 60 inches long [14]. This furnace is mounted in a 15,000-psi pressure vessel with a cavity of 27 inches ID by 108 inches long. The KBI pressure vessel, measuring 36 inches

Figure 16. Hot isostatic pressure vessel designed by ASEA, utilizing wire-wound frame. (Courtesy ASEA, Inc.)

ID, will have a 27-inch ID furnace capable of operation at temperatures in excess of 2000°F. This will be the largest facility currently in operation.

It appears that size capability can be significantly increased within the current state of the art for pressure vessels. Within the past year, Battelle performed an in-house engineering study into the feasibility of building a facility that included a HIP pressure vessel of 5 feet ID by 14 feet inside length operated at 15,000 psi. Four of the major pressure-vessel manufacturers responded positively to our inquiries with preliminary quotations on the cost of construction of this vessel. Similarly, we found that compressors could be built to pressurize this system within a reasonable length of time. Our conclusion was that such a facility is currently well within the state of the art. We would estimate that a pressure vessel of this type could be operated at 1900°F with a furnace in excess of 42 inches ID, or at 2400°F with a furnace of approximately 36 inches ID.

Summary

It has been clearly demonstrated that the HIP process is capable of fabricating unusual shapes. Through the use of this approach, it is conceivable that many structures could be pressed to within 80-90 percent of net configuration. Coupling this approach with secondary fabrication, the possibility exists that intermediate steps may be eliminated through the use of HIP preforms.

The use of HIP net shapes or preforms becomes more attractive when this ability is coupled with the relatively high levels of properties that can be achieved in certain HIP materials. One must be impressed with the potential of making a titanium alloy shape with full wrought properties. Undoubtedly, the area of achieving wrought properties in a P/M product is one that needs more fundamental work. The scientific community is better equipped today than ever before with the capability to quantify the surfaces of powder particles through the use of tools such as scanning electron microscopy, sputtering mass spectrometers, and laser microprobes.

Equipment has been one of the limiting factors in determining rate and size of production items. HIP has been regarded in the past as a slow process, limited to relatively small sizes. As a result, the process has been considered by most people as an expensive way to make P/M parts. However, the rate at which parts can be handled through a pressure vessel, improvements in canning techniques, and the high quality of the resultant product have all contributed to making HIP an economically competitive powder-metallurgy process.

Acknowledgments

I wish to express my gratitude to the equipment and materials manufacturers mentioned in the text who were kind enough to supply data and photographs. Also, I would like to acknowledge the contribution of various Battelle staff members who have contributed significantly to the state of the art of HIP, including Messrs. C. B. Boyer, F. D. Orcutt, P. J. Gripshover, J. J. Mueller, G. H. Harth, A. N. Ashurst, and J. H. Peterson.

References

1. Hanes, H.D., Gripshover, P.H. and Hodge, E.S., "Fabrication of Complex Beryllium Shapes by Gas-Pressure Compaction," *Proceedings of the Conference Internationale sur la Metallurgie du Beryllium*, Paris: Presses Universitaires de France (1965), 579.
2. Gripshover, P.J. and Hanes, H.D., "Advanced Beryllium Gyro-Materials Technology," 5th Space Congress, Cocoa Beach, Florida, March 1968.

3. Ashurst, A.N., Goldstein, M. and Ryan, M.J., "Techniques for Fabricating Rocket Engine Components Containing Intricate Flow Channels," Battelle Memorial Institute Metal Science Group, Columbus, Ohio, NASA Contract Report, NASA–CR–72588 Final Report, 1 August 1969 (N69–36897).

4. Johnson, M.R., Hanes, H.D. and Pinkerton, G.B., "Hot Isostatic Pressing of Large, Hollow, Structural Beryllium Shapes," ASTM Annual Meeting, San Francisco, California, June 1968.

5. Mueller, J.J. and Hanes, H.D., "Establishment of a Manufacturing Process for Thin Walled Conical Beryllium Structures Involving Hot Isostatic Pressing," Phase I, Battelle Memorial Institute, Columbus Laboratories, Columbus, Ohio, Air Force Materials Laboratory Contract Report RTD–IR–271–9 (I), April 1970 (A1873 271).

6. Mueller, J.J. and Hanes, H.D., "Establishment of a Manufacturing Process for Thin Walled Conical Beryllium Structures Involving Hot Isostatic Pressing," Phase II, Battelle Memorial Institute, Columbus Laboratories, Columbus, Ohio, Air Force Materials Laboratory Contract Report RTD–IR–271–9 (I), April 1970 (A1873 271).

7. Lidman, W.G. and Younger, F.K., "Hot Isostatic Pressing of Beryllium," Proceedings of the Beryllium Conference, National Materials Advisory Board Report, NMAB–272, July 1970 (AD 710–704).

8. Harth, G.H., Personal communication.

9. Harth, G. H., "New Process for Titanium P/M" *Precis. Metal*, 28 (1970), 43.

10. Sajdak, R.J. and Geisendorfer, R., Personal communication.

11. "P/M Process Improves High-Speed Tool Steel," *Prod. Eng.*, 42 (1971), 55.

12. Hellman, P. *et. al.*, "The ASEA-Stora Process: A New Process for the Manufacture of Tool Steel and Other Alloy Steels from Powder," International Powder Metallurgy Conference, New York, July 1970, in *Modern Developments in Powder Metallurgy,* Vol. 4, *Processes,* Henry H. Hausner, ed., New York: Plenum Press (1971), 573.

13. "Sandvik Steel Now Producing Tungsten Carbide by Hot Isostatic Compaction in New ASEA Equipment," *ASEA News*, August 1971.

14. Boyer, C.B., Orcutt, F.C. and Hatfield, J.E., "Hot Isostatic Bonding and Compaction Development," *Ind. Heat.,* 37 (1970), 50.

12. Fabrication of High-Strength Aluminum Products from Powder

J. P. LYLE, JR. and W. S. CEBULAK
Alcoa Technical Center
Pittsburgh, Pennsylvania

ABSTRACT

Better combinations of strength, resistance to stress-corrosion cracking, and resistance to exfoliation can be obtained in extrusions and forgings made from prealloyed atomized powder than in corresponding products made from ingot. Internal quality meeting ultrasonic SNT Class A Airframe standards can be consistently obtained by a process consisting of cold compacting, preheating, hot pressing, hot working, solution heat-treating, quenching, and artificial aging. Decreasing powder size tends to increase transverse ductility. Green density has no effect on properties, but 70–80 percent of theoretical density is optimum for processing. The pressure required to obtain this density varies from 15 to 65 ksi and depends on compacting method, alloy content, and time elapsed between atomizing and compacting. Preheating for about one hour at a temperature at least as high as the solution heat-treatment temperature is required to prevent porosity, blistering, and delamination during solution heat treatment of the final product, and to develop maximum ductility. Hot compacting pressure has no effect on properties, but quality of hand forgings does improve with increasing pressure. Fracture toughness increases with increasing amounts of hot deformation.

Introduction

Better combinations of strength, resistance to stress-corrosion cracking (SCC), and resistance to exfoliation corrosion can be obtained with products fabricated from aluminum alloy powder than from corresponding products made from ingot. An early example of combined high strength and high resistance to SCC is shown in Figure 1. A 7075-T6 extrusion made from atomized powder had strength in the same range as an extrusion and a rolled and drawn rod made from ingot; P/M (powder metallurgy) extrusions survived one year in 3.5-percent NaCl

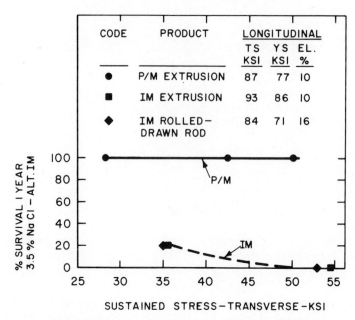

Figure 1. Tensile and stress-corrosion properties of 7075 alloy 0.75-inch-diameter rod.

alternate immersion at sustained stresses of 35–50 ksi in the transverse direction, while only 0–20 percent of the I/M (ingot metallurgy) products survived under the same conditions.

Although the possibility of improved properties was demonstrated many years ago, quality of high-strength aluminum alloy products made from powders was erratic. In particular, blistering was encountered during solution heat-treatment; on numerous occasions, ductility in the transverse direction was poor. Substantial improvements in quality were made by 1962 when it was reported that extrusions of high ultrasonic quality (exceeding SNT Class A Airframe requirements) could be made consistently [1]. Extrusions made by the improved techniques also were capable of better combinations of strength, resistance to SCC, and resistance to exfoliation than either commercial or experimental alloys made from ingot [2]. Some of the combinations of strength and resistance to SCC which were reported for P/M extrusions by Haarr [2] are compared in Table I with those of two tempers of 7075 extrusions made from ingot. Some tempers of P/M alloys E, D, and J had higher strengths than I/M 7075-T6, especially in the transverse direction. The P/M alloys also possessed remarkably superior resistance to SCC as measured by the alternate immersion test, equal to that of I/M 7075-T73 extrusions.

The SCC tests reported above were made in the transverse direction through the center line of the rod. A more critical SCC situation exists in the short-

TABLE I

Tensile and Stress-Corrosion Properties of Some Aluminum Alloys[1]

Alloy[4]		Age (Hrs at °F)	Longitudinal			Transverse			SCC[2] (ksi)
Type	No.		T.S. (ksi)	0.2% Y.S. (ksi)	El. in 4D (%)	T.S. (ksi)	0.2% Y.S. (ksi)	El. in 4D (%)	
P/M	D	24 at 250	99	93	5	87	81	2	41[3]
P/M	J	24 at 250 + 3 at 330	94	88	7	84	79	3	40[3]
I/M	7075	−T6	92	85	10	77	67	7	15
P/M	E	6 at 250 + 8 at 330	90	81	7	84	76	3	57[3]
P/M	D	6 at 250 + 8 at 330	84	78	8	77	71	4	53[3]
I/M	7075	−T73	78	70	12	71	60	7	48

Notes: 1. 2″-diameter extrusions.
2. Highest stress at which no failures occurred in 84 days in 3.5% NaCl solution in alternate immersion. 0.125″ tensile bars.
3. Reference [2].
4. See Table IV for compositions.

transverse direction across the parting plane of some die forgings which have a highly worked and aligned structure. P/M alloys show promise for this kind of application also, as can be seen in Table II. Alloy H made from powder had practically the same strength as 7075 made from ingot, and substantially superior resistance to SCC. Alloy 7075 made from powder had better resistance to SCC than its counterpart made from ingot, but the strength of the P/M version was lower. This illustrates also that a simple change from ingot metallurgy to powder metallurgy may not give optimum combination of properties, but that changes in composition also must be made.

TABLE II

Tensile, Toughness, and Stress-Corrosion Properties of a Small Die Forging[1]

Alloy[5]		Longitudinal				Short-Transverse[2]			SCC[4]
Type	No.	T.S. (ksi)	0.2% Y.S. (ksi)	El. in 4D (%)	UPE[3]	T.S. (ksi)	0.2% Y.S. (ksi)	El. in 4D (%)	
P/M	H	82	75	9	120	78	71	4	100% Pass
P/M	7075	80	72	11	45	76	68	7	67% Pass
I/M	7075	82	78	11	195	80	73	5	0% Pass

Notes: 1. Door bracket, Part No. 9–56401, aged 24 hrs at 250°F + 8 hrs at 330°F.
2. Across flash plane.
3. Unit propagation energy (in.–lb/in.^{2}).
4. Stress-corrosion cracking: 84 days in 3.5% NaCl alternate immersion, 0.125″ tensile bars stressed at 42 ksi.
5. See Table IV for compositions.

Toughness is a shortcoming of P/M alloys which must be overcome, as illustrated in Table II. Toughness (as measured by the unit propagation energy) of both P/M alloys is lower than that of the I/M alloy.

P/M products have higher resistance to exfoliation corrosion at high strength levels than I/M products. In Table III, P/M alloy H and I/M alloy 7175-T66 have

TABLE III

Tensile and Exfoliation Properties of a Small Die Forging[1]

Alloy[3]		Longitudinal			Transverse			Resistance to Exfoliation[2]
Type	No.	T.S. (ksi)	0.2% Y.S. (ksi)	El. in 4D (%)	T.S. (ksi)	0.2% Y.S. (ksi)	El. in 4D (%)	
P/M	H	90	84	7	88	81	4	Excellent
I/M	7175–T66	91	86	10	90	84	10	Poor
I/M	7075–T6	86	77	11	83	73	9	Poor
I/M	7175–T736	77	68	14	76	68	9	Excellent

Notes: 1. Part No. 62510.
2. Based on results of EXCO Test. Total immersion in 4N NaCl + 0.5N KNO_3 + 0.1N HNO_3 (pH = 0.4) for at least 72 hours. See Reference [6] for method.
3. See Table IV for compositions.

practically the same strengths, but resistance of the P/M material to exfoliation is superior. Alloy H is superior to I/M 7075-T6 in both strength and resistance to exfoliation. Exfoliation resistance of the I/M alloy can be improved by a change in temper, e.g., 7175-T736, but the change produces lower strength.

High strengths, resistance to stress-corrosion cracking, and resistance to exfoliation of P/M products relative to I/M products are very desirable. These properties are influenced by the fabrication processes. The purpose of this paper is to describe the important steps in fabrication of P/M products and to illustrate the effects those steps have on subsequent fabricating operations and on properties of the final product.

Fabricating Process

Important fabricating steps are: (1) cold compacting; (2) preheating; (3) hot compacting; (4) scalping; (5) hot working; (6) solution heat-treating and aging.

Prealloyed atomized powder is the starting material for the fabricating process. Some compositions of interest are listed in Table IV.

The primary purpose of atomizing is to achieve a billet with much finer solidification metallurgical structure than can be obtained in ingot. These size effects are very important and influence subsequent fabricating operations as well as the properties of the final product.

TABLE IV

Compositions of Some Powder-Metallurgy and Ingot-Metallurgy Alloys

Alloy No.	Former No.[1]	Weight, %				Comments
		Zn	Mg	Cu	Others	
A	–	5.9	2.6	1.8	0.8Fe–0.8Ni	P/M and I/M
B	–	6.5	2.3	1.5		P/M
C	–	6.6	2.3	2.3	0.11Zr	P/M variant of X7050
D	90	7.5	2.4	1.0	1.1Fe–1.0Ni–0.2Cr	P/M
E	87	7.6	2.5	1.1	2.2Fe–2.3Ni–0.2Cr	P/M
F	–	8.0	2.5	1.0		P/M
G	–	8.0	2.5	1.0	1.6Co	P/M
H	–	8.0	2.5	1.0	0.8Fe–0.8Ni	P/M
J	71	9.0	3.5	0.5	0.75Co	P/M
7075		5.6	2.5	1.6	0.3Cr, 0.50Fe max	I/M and P/M in this chapter
7175		5.6	2.5	1.6	0.25Cr, 0.20Fe max	I/M
X7050		6.1	2.3	2.4	0.12Zr	I/M

Note: 1. References [1,2,4].

Powder particle-size distribution depends primarily on atomizing conditions, although minor changes may be made by screening. Changes through classification generally are not followed because of inefficiency and cost. Examples of powder size distributions from screen and micromesh sieve analyses are given in Figure 2 for powders of interest.

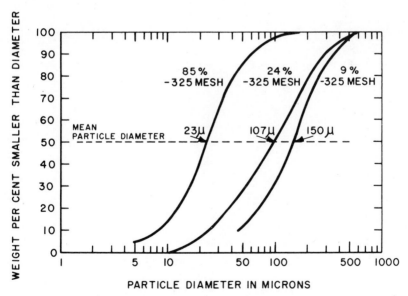

Figure 2. Particle-size distribution for three atomized powders.

Structures of ingot and powder of alloy A are compared in Figure 3. Both eutectic and larger primary $FeNiAl_9$ particles are clearly seen in the ingot samples along cell boundaries. Cells are much finer in the powders than in the ingot and $FeNiAl_9$ particles are not discernible, even though their presence is indicated by X-ray diffraction. From the relationship between cell size and solidification rate shown by Matyja, Giessen, and Grant [5], the derived solidification rates are 9.5×10^4 and 7×10^3 °F/sec for 25 μ and 115 μ powders, respectively, and 30 and 3 °F/sec for 6-inch- and 16-inch-diameter ingots, respectively.

The relationship between ingot or particle size and the as-cast cell size for alloy A is plotted in Figure 4. This figure also shows that the effects of the dif-

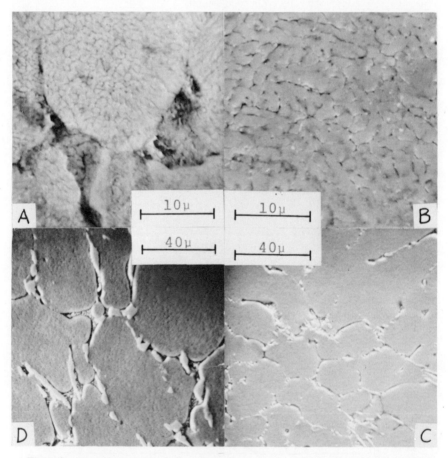

Figure 3. As-cast structures of alloy A atomized powders and ingots. (A) 25 μ powder, average cell size = 0.9 μ. (B) 115 μ powder, average cell size = 1.8 μ. (C) 6″ ingot, average cell size = 15 μ. (D) 16″ ingot, average cell size = 42 μ.

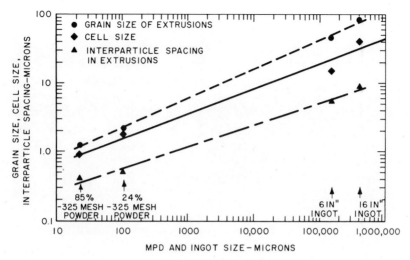

Figure 4. Effect of powder particle size and ingot size on the structure of alloy A.

ferences in as-cast structure can be seen in structures of the final products. For example, grain size of the extrusions shows the same trends as cell size of the original materials. Spacing between insoluble constituent particles, FeNiAl$_9$ in this example, in the extrusions shows the same trend. Microstructures of extrusions are shown in Figure 5.

Specific surface area of powder increases with decreasing particle size, resulting in an increase in oxygen in the fabricated product (Figure 6). In alloys containing Mg, this is usually in the form of MgO. While effects of MgO have not been studied, it is expected to retard grain growth, thereby improving both ductility and toughness. As an insoluble constituent, however, it may have the opposite effect.

In alloys which contain substantial amounts of insoluble constituent, elongation frequently improves with increasing fineness of structure, especially in the transverse direction. This is illustrated in Table V with alloy A, where elongation increases from 6 to 10 percent as grain size decreases from 85 to 2 microns.

Another important effect of increasing fineness of metallurgical structure is an increase in quench sensitivity in alloys which do not contain chromium. In the absence of $Al_{12}Mg_2Cr$ phase, grain boundaries and insoluble constituent particles are very effective as precipitation sites and as lattice-vacancy sinks. Consequently, quench sensitivity increases with decreasing grain size and decreasing distance between insoluble constituent particles, as shown in Table VI. A corresponding trend is observed with X7050–Alloy C composition in Figure 7. It is important to note that alloy C is less quench-sensitive than I/M 7075.

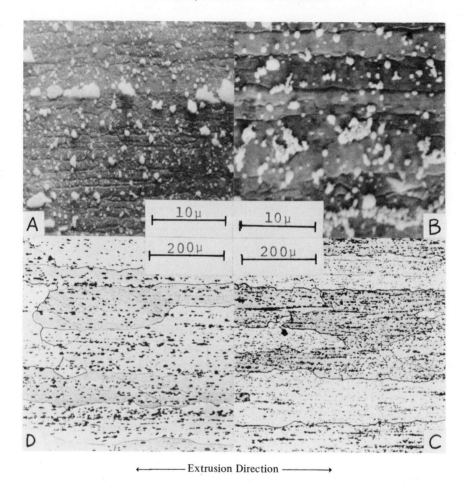

←——————— Extrusion Direction ———————→

Figure 5. Structures of alloy A extrusions made from powders and ingot (extrusion ratio = 8). (A) 23 μ powder: grain size = 1.25 μ; interparticle spacing = 0.4 μ. (B) 107 μ powder: grain size = 2.25 μ; interparticle spacing = 0.5 μ. (C) 6″ ingot: grain size = 46.3 μ; interparticle spacing = 5.4 μ. (D) 16″ ingot: grain size = 85 μ; interparticle spacing = 8.9 μ.

Cold Compacting

Cold compacting is a preliminary step to make subsequent preheating and hot compacting operations more convenient. Both uniaxial and isostatic compacting have been used successfully.

Tensile properties and NTS/YS of extrusions and forgings are not affected by green density of the compacts from which they are made, nor by the compacting method. Any original differences are eliminated by hot working.

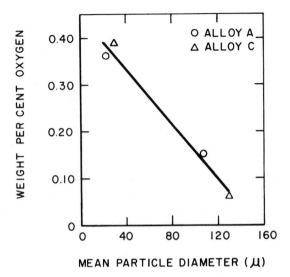

Figure 6. Effect of powder size on oxygen content of argon-preheated extrusions.

Green densities are important in subsequent processing. Probability of chipping or breaking of compacts during preheating and hot pressing decreases with increasing green density. If hot pressing is done in an extrusion cylinder just before extrusion, scrap losses increase with decreasing green density because a larger fraction of the compact must be left in the extrusion butt.

The costs of cold compacting and hot compacting tools are interrelated. Cold compacting tool cost increases with increasing green density due to higher compacting pressures and the increased cylinder wall thickness required. Cost of hot

TABLE V

Effect of Structure on Tensile Properties of Alloy A Extrusions

Ingot Dia. (in.)	Powder MPD (μ)	Grain Size (μ)	Inter-particle Spacing (μ)	T.S. (ksi)	0.2% Y.S. (ksi)	El. in 4D (%)
			Extrusion[1]			
—	23	1.25	0.4	83	74	10
—	107	2.25	0.5	87	78	10
6	—	46.3	5.4	85	74	8
16	—	85.0	8.9	78	71	6

Notes: 1. 1" X 4.25" round-end bar; extrusion ratio = 8. Solution heat-treated 2 hrs at 900°F, quenched in cold water, aged 24 hrs at 250°F + 4 hrs at 350°F.

TABLE VI

Effect of Structure on Quench Sensitivity of Alloy A Extrusions

| Ingot Size (in.) | Powder MPD[4] (μ) | Cell Size (μ) | Grain Size (μ) | Inter-particle Spacing (μ) | Extrusion[1] | | Y.S. (BWQ) Y.S. (CWQ) |
| | | | | | Transverse Y.S. (ksi) | | |
					CWQ[2]	BWQ[3]	
–	23	0.9	1.25	0.4	76	48	0.63
–	117	1.8	2.25	0.5	79	56	0.71
6	–	15.2	46.3	5.4	74	69	0.93
16	–	42.0	85.0	8.9	69	69	1.00

Notes: 1. 1″ × 4.25″ Bar; extrusion ratio = 8. Solution heat-treated 2 hrs at 900°F, quenched, aged 24 hrs at 250°F + 2 hrs at 350°F.
2. Cold-water quench. Average cooling rate = 150°F/sec from 750–550°F.
3. Boiling-water quench. Average cooling rate = 3°F/sec from 750–550°F.
4. Mean particle diameter.

compacting tools decreases with increasing green density because they may be smaller to accommodate the same weight of metal.

In practice, optimum green density is 70–80 percent of theoretical density. Below 65 percent, danger of chipping and breaking is high. Above 95 percent, compacting pressures are excessive.

Lubricant is not mixed with the powder because of the danger that residues will cause blistering during solution heat treatment of the fabricated product or

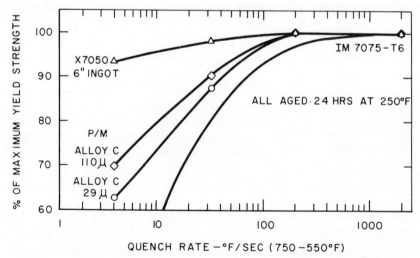

Figure 7. Effect of structure on quench sensitivity of alloy C, X7050 and 7075 [7,8].

that low ductilities will be obtained in the final product. For uniaxial compacting, a lubricant (usually butyl stearate) must be applied to the die surfaces. A tapered die cavity greatly facilitates ejection of the compact without cracking. The dies shown schematically in Figure 8 have been used with very satisfactory results, provided that the dummy-block face is at the bottom of the straight portion of the die at the finish of the compacting stroke.

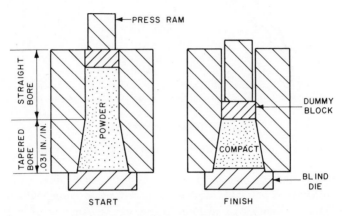

Figure 8. Uniaxial compacting tools [1].

Pressures required to achieve a given density depend on compacting method, on alloy composition, and on natural aging which occurs between atomizing and cold compacting. Isostatic compacting gives higher density for a given pressure than does uniaxial compacting. More highly alloyed compositions require higher pressures (Figure 9). Because substantial Zn, Mg, and Cu are retained in supersaturated solid solution by the rapid quench during atomizing, alloy particles will harden during storage at room temperature. As a result, higher compacting pressures are required to reach a given density as the storage time increases (Table VII).

Preheating

The primary effect of preheating is to lower gas content of the compact so as to prevent subsequent blistering and delamination in the final product during solution heat-treatment. There are also important changes in structure which affect properties of the final product.

Gas present in the final worked product made from Al–Zn–Mg–Cu alloy powders is predominantly hydrogen, originating as adsorbed water and hydrates

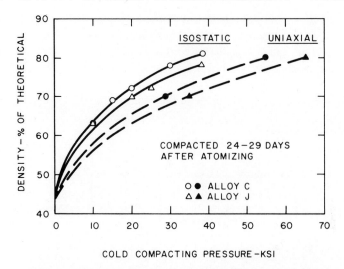

Figure 9. Effect of alloy and compacting method on density [3].

on the powder surfaces picked up during atomizing and subsequent handling. When the powder is reheated, some water is evolved as vapor, but some reacts with aluminum or magnesium to form hydrogen. The amount of hydrogen remaining after preheating depends primarily on preheat temperature. Blistering, porosity, and delamination in the final product can be prevented by preheating the porous compact to a temperature which is approximately as high as any temperature that the compact or fabricated product will subsequently en-

TABLE VII

Effect of Natural Aging Time on Compacting Pressure

Time[1] (days)	Compacting Pressure[2] (ksi)	Density, % of Theoretical	Alloy
2	25.5	70	C
24	28.6	70	C
2	47.4	80	C
24	54.8	80	C
4	32.2	70	J
25	34.8	70	J
5	60.1	80	J
25	65.1	80	J

Notes: 1. Time between atomizing and compacting.
2. Pressure required to reach indicated density (6"-diameter uniaxial compacts).
3. From Reference [3].

counter, usually the solution heat-treating temperature. Preheating is conducted in a flowing, nonoxidizing gas. Degassing is greatly facilitated by the fact that porosity of the compact is largely interconnected.

An early example of the benefit of preheating is shown in Figure 10.

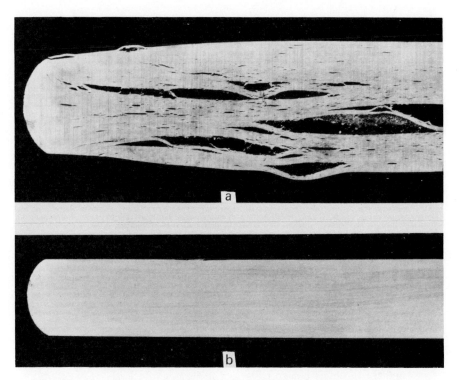

Figure 10. Macrosections from P/M 7075-W forged biscuits. Delamination occurred during solution heat-treatment at 860°F. (a) No preheat; (b) Preheated 24 hrs at 960°F. 1X.

Effects of preheat temperature on hydrogen content and transverse elongation of some extrusions are shown in Figure 11. Hydrogen content decreases rapidly to about 3 ml of $H_2/100$ g of Al at about 950 °F, falling slowly thereafter. Transverse elongation rises rapidly to about 900 °F, peaks near 950 °F, then falls. Solution heat-treatment temperature was 860 °F, and it is apparent that best transverse elongation was obtained by preheating above that temperature.

The fall in elongation at the higher preheat temperature is at least partly the result of extreme coarsening of insoluble constituent. $Al_{12}Mg_2Cr$ constituent particles approaching 30 microns in diameter were reported in the extrusion from the compact preheated at 1100°F.

Additional evidence of coarsening of a constituent having low solubility at

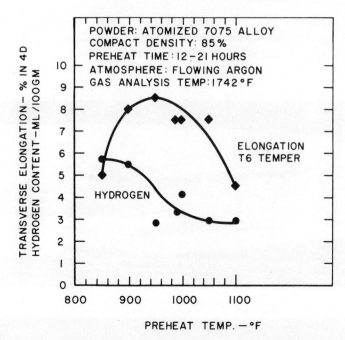

Figure 11. Effect of compact preheat temperature on hydrogen content and transverse elongation of extrusions [1].

preheat temperatures is given in Table VIII. Growth of Co_2Al_9 is accompanied by a decrease in toughness as indicated by the ratio of notched-tensile strength to yield strength (Figure 12).

Because of the complex interaction of several factors, including gas content, coarsening of insoluble constituents, and melting, over-all effects of preheating times and temperatures on properties are difficult to predict quantitatively. Effects on tensile strength and yield strength are small in the 900–1000°F range for times up to at least 20 hours (Table IX). More pronounced effects are seen on elongation and NTS/YS in the transverse direction, as shown in Figure 13 for alloy J and Figure 14 for alloy C. In general, 900°F for one hour gives about the best over-all transverse NTS/YS and elongation.

Hot Compacting

The purpose of hot compacting is to raise the hot ductility of the compact sufficiently to permit hot working. When the compact is to be extruded with a high extrusion ratio, a separate hot compacting operation may not be required. When hot compacting is used prior to extrusion, it is usually done in the extru-

TABLE VIII

Effect of Preheat Temperature and Time on the Size of
Co_2Al_9 Particles in Alloy J Forgings[1]

Preheat Temp. (°F)	Preheat Time (hrs)	Average Co_2Al_9 Size (μ)
900	1	0.40
900	5	0.65
900	20	0.80
1000	1	0.65
1000	5	1.10
1000	20	1.50

Note: 1. Compacts preheated in flowing argon, hot
compacted at 90 ksi, forged to 2″ × 2″ bar,
solution heat-treated 2 hrs at 920°F,
quenched in cold water (from Reference
[3]).

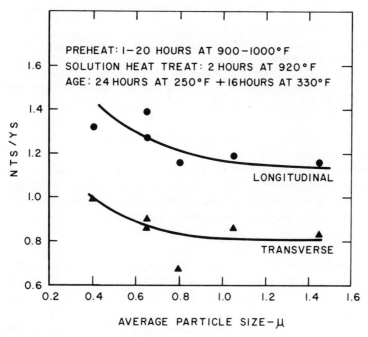

Figure 12. Effect of Co_2Al_9 particle size on the toughness of alloy J forged bar [3].

TABLE IX

Effect of Preheat Time and Temperature on the Tensile Properties of Forged Bar[1]

Preheat Temp. (°F)	T.S. (ksi)			0.2% Y.S. (ksi)			Elongation (% in 4D)			NTS/YS		
	1 Hr	5 Hrs	20 Hrs	1 Hr	5 Hrs	20 Hrs	1 Hr	5 Hrs	20 Hrs	1 Hr	5 Hrs	20 Hrs
Alloy J—Longitudinal Direction												
900	74	74	74	65	63	64	15	16	15	1.29	1.23	1.20
950	–	74	–	–	64	–	–	16	–	–	1.24	–
1000	74	75	76	65	66	66	16	16	14	1.27	1.21	1.14
Alloy J—Transverse Direction												
900	72	72	72	61	61	62	10	8	9	0.93	0.84	0.57
950	–	70	–	–	61	–	–	6	–	–	0.71	–
1000	73	73	73	62	63	63	9	11	9	0.79	0.81	0.74
Alloy C—Longitudinal Direction												
900	76	77	75	68	68	67	15	14	16	1.43	1.34	1.29
950	77	74	77	69	67	69	15	17	14	1.35	1.35	1.24
1000	75	74	75	68	67	68	16	17	17	1.40	1.35	1.34
Alloy C—Transverse Direction												
900	74	70	75	65	62	65	12	6	7	0.92	0.65	0.71
950	75	72	70	65	64	63	11	9	6	0.93	0.88	0.58
1000	72	71	72	62	63	63	10	8	12	0.86	0.90	0.83

Notes: 1. 2" × 2" bar, forged from 6"-diameter compacts. Green Density: 70–80%; preheated in flowing argon; hot pressed at 60–90 ksi; alloy J was solution heat-treated 2 hrs at 920°F, CWQ, aged 24 hrs at 250°F + 16 hrs at 330°F; alloy C solution heat-treated 2 hrs at 890°F, CWQ, aged 24 hrs at 250°F + 8 hrs at 330°F (from Reference [3]).

sion cylinder against a blind die. The hot pressed compact is left in the cylinder, the blind die replaced with an extrusion die, and extrusion proceeds. When a compact is to be forged without an intermediate extrusion operation, a separate hot compacting operation is required to give sufficient ductility for forging.

Hot pressing immediately follows preheating to prevent regassing and oxidation of the porous compact. Hot pressing and preheat temperatures are therefore the same.

If the hot compact is to be ejected from the die, it is important to use a die cavity similar to that used for cold uniaxial compacting (Figure 8). In this case, density of the compact must be essentially 100 percent of theoretical alloy density when the face of the dummy block is at the bottom of the straight portion of the die.

Density of the compact is raised to essentially 100 percent in the hot compacting operation at pressures as low as 30 ksi. Forgeability, however, increases

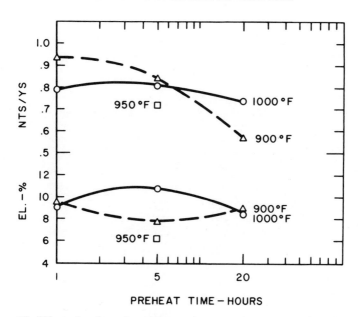

Figure 13. Effect of preheat time and temperature on transverse NTS/YS and elongation of alloy J forged bar [3].

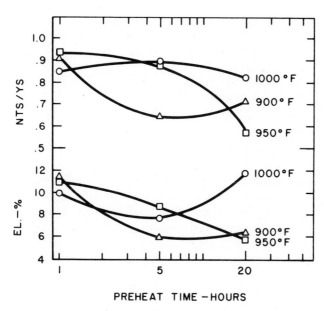

Figure 14. Effect of preheat time and temperature on transverse NTS/YS and elongation of alloy C forged bar [3].

with increasing hot compacting pressures above 30 ksi. This was evident in recovery, or the percentage of the finished forgings which met SNT Class A quality standards (Table X).

TABLE X

Effect of Hot Compact Pressure on Quality of Forgings[1]

Alloy	Green Density[2] (%)	Preheat[3]		Quality[4] Hot Compact Pressure		
		Time (Hrs)	Temp. (°F)	30 (ksi)	60 (ksi)	90 (ksi)
C	70	5	950	88	87	92
C	80	5	900	82	87	86
C	80	5	950	65	91	100
C	80	5	1000	88	90	95
C	70[5]	5	950	28	92	93
J	70	5	950	25	70	93
J	80	5	950	10	78	71
J	80	5	1000	38	68	79
J	70[5]	5	950	76	89	91
			Average	56	84	89

Notes: 1. 2″ × 2″ bar, forged from 6″-diameter compacts.
2. Uniaxial compact except as noted.
3. In flowing argon.
4. Percent of forging which met SNT Class A.
5. Isostatic compact.
6. From Reference [3].

Tensile properties of forged bar are not significantly affected by the hot compacting pressure at which the compact is made.

Scalping

Some surface must be removed from the hot pressed compact to prevent cracking during forging. The amount of metal removed from the radius is approximately $\frac{1}{8}$ inch. The amount which must be removed from the ends of the compacts apparently depends on the length–diameter ratio of the compact. It is also apparent that less metal must be removed from the "ram end" than from the "blind-die end" of the hot pressed compact.

Hot Working

Compacts are hot worked at 550–700°F.
Hot deformation in which substantial metal flow occurs is required to raise

the ductility of products made from atomized Al–Zn–Mg–Cu alloy powders. Compacts, even when preheated and hot pressed by good practices, have low ductilities in comparison with extrusions and forgings (Table XI).

TABLE XI

Effect of Hot Working on Properties of P/M 7075–T6

Form Tested	Preheat			Hot Compacting Pressure (ksi)	Density (%)	Longitudinal			Transverse[5]		
	Atmo-sphere	Temp. ($^{\circ}$F)	Time (Hrs)			T.S. (ksi)	0.2% Y.S. (ksi)	El. in 4D (%)	T.S. (ksi)	0.2% Y.S. (ksi)	El. in 4D (%)
Extrusion[1]	Argon	990	21	93	100	92	83	10	80	72	6
Compact[2]	Argon	990	21	75	100	–	–	–	80	74	1
Forging[3]	Vacuum	925	4	93	100	86	71	12	–	–	–
Compact[4]	Vacuum	925	4	93	100	68	65	2	–	–	–

Notes: 1. 2″-diameter rod; extrusion ratio = 10 (from Reference [4]).
2. 2.31″ diameter.
3. 3/8″ rod, forged from a 2″ × 2″ piece of the compact for which properties are reported. Reduction ratio = 36.
4. 6″ diameter.
5. Along the diameter of the extruded rod or along the axis of the compact, parallel to the direction of ram movement.

Increasing amounts of hot deformation tend to increase NTS/YS and transverse A elongation and to reduce property anisotropy (Figure 15). This effect is particularly significant in the 80- to 95-percent reduction range (L = 5–20).

Solution Heat-Treating and Aging

Principles of solution heat treating P/M Al–Zn–Mg–Cu alloys are the same as for ingot alloys. Best combinations of properties are obtained by holding sufficiently long at the highest temperature possible without melting to put all soluble elements into solution. Figure 16 summarizes the effects of composition on solution heat-treatment temperatures as derived from X-ray, metallographic, or tensile tests.

Soak times at solution heat-treatment temperatures can be shorter for P/M products than for I/M products because of the finer structure of the former. This effect can be shown by the length of time which is required to put all of the S-phase (Al_2CuMg) into solution (Table XII).

Strengths of both P/M and I/M Al–Zn–Mg–Cu alloys decrease with decreasing quench rates (Figure 7 and Table VI). Cr-free P/M alloys are more quench-

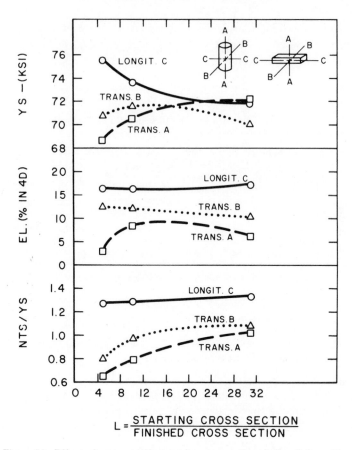

Figure 15. Effect of amount of hot work on properties of alloy B forged bar.

sensitive than their I/M alloy counterparts, but less so than I/M Al-Zn-Mg-Cu alloys containing Cr.

The aging of P/M alloys is similar to that of I/M alloys. The stress-corrosion resistance is usually improved by a two-step age, the first being in the 225-275°F range and the second in that of 325-350°F. Strengths characteristically fall during the second step. Aging practices are developed empirically for each alloy and product to give optimum combination of properties.

Status

Based on the principles discussed above, cold and hot compacting tools have been designed and built to produce compacts 8.4-9.2" in diameter × 28" long, weighing 170 pounds. Compacts have been produced in alloys B, F, and G. Hot

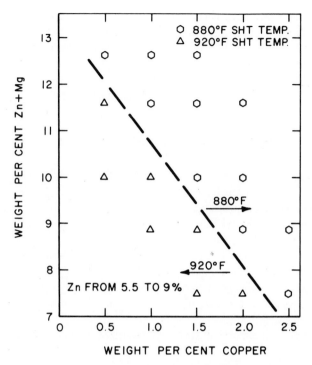

Figure 16. Solution heat-treatment temperatures for P/M Al–Zn–Mg–Cu alloys [8].

TABLE XII

Time Required to Dissolve S-Phase in Hand-Forgings[4]

Time (Hrs)	Temp. (°F)	Amount of S-Phase[1]	
		Alloy C[2]	X7050[3]
1	888	None	Medium
4	888	None	Medium
1	907	None	Medium −
4	907	None	Small

Notes: 1. Amount of Al_2CuMg estimated by X-ray diffraction using a Guinier-DeWolfe focusing camera and comparing patterns on films to patterns of known phase amounts.
2. Made from 85% 325-mesh powder; cell size = 2.4 μ.
3. Made from 28″-diameter ingot; cell size = 66 μ.
4. Preheat 2 hrs at 890°F. Upset 67% to biscuit, solution heat-treated, CWQ.

pressed compacts have been hand-forged to square and rectangular bar for property surveys (Figure 17) and for further fabrication to die forgings. Rectangular forged slab has been hot rolled to plate for property surveys. Plans call for hot and cold rolling to sheet. Large compacts have been extruded to shapes for property evaluation, and to rod which was die-forged to a rib-and-web configuration (Figure 18), which will be used for a critical evaluation of resistance to SCC.

Summary

Products having high internal quality can be fabricated from atomized Al–Zn–Mg–Cu alloy powders by a process consisting of cold compacting, preheating, hot pressing, and hot working.

Ductility in the transverse direction of the product tends to increase with decreasing powder size, but quench sensitivity also increases.

Green density has no effect on properties of the final wrought product, but 70–80 percent of theoretical density is the optimum for processing. Pressure required to reach a desirable green density in this range varies from 15 to 65 ksi depending on compacting method, alloy content, and time elapsed between atomizing and compacting.

Preheating at temperatures at least equal to the solution heat-treatment tem-

Figure 17. Alloy B hand-forging and compacted forging stock. Left rear: hot pressed compact, 8.4–9.2″ diameter X 28″ long. Right rear: scalped compact, 7.5″ diameter X 19″ long. Front: 5″ X 5″–3″ X 3″ stepped hand forging.

Figure 18. Production sequence for alloy B die forging. (a) Die forging; (b) extruded rod stock; (c) hot pressed compact; (d) cold compact.

peratures is required to prevent porosity, blistering, and delamination during solution heat-treatment of the final wrought product and to develop maximum ductility in the product.

Hot compacting pressure has no effect on properties but increases the quality of forgings.

Hot deformation is required to develop ductility, with properties increasing with increasing amounts of hot deformation.

P/M products require less time for solution heat-treatment than I/M products.

Acknowledgments

This work is supported by The Aluminum Company of America and by a Manufacturing Methods Contract with the U.S. Army, administered by Messrs. Harold Markus and Donald H. Kleppinger at Frankford Arsenal and funded jointly by the Aviation Systems Command (AVSCOM), St. Louis, Missouri; the Munitions Command (MUCOM), Dover, New Jersey; and the Production Equipment Agency (PEQUA), Rock Island, Illinois.

References

1. Towner, R.J., "Development of Aluminum Base Alloys," Alcoa Research Laboratories, New Kensington, Pennsylvania, Frankford Arsenal Contract Report 7–62–AP59S, Annual Progress Report, September 1961–62, October 1962.
2. Haarr, A.P., "Development of Aluminum Base Alloys–Section III," Alcoa Research Laboratories, New Kensington, Pennsylvania, Frankford Arsenal Contract Report 13–65–AP59S, May 1966 (AD 487–764).
3. Cebulak, W.S. and Truax, D.J., "Program to Develop High-Strength Aluminum Powder Metallurgy Products: Phase I–Process Optimization," Alcoa Research Laboratories, New Kensington, Pennsylvania, Frankford Arsenal Contract Report, March 1971 (AD 882–137 L).
4. Haarr, A.P., "Development of Aluminum Base Alloys–Section II," Alcoa Research Laboratories, New Kensington, Pennsylvania, Frankford Arsenal Contract Report 13–65–AP59S, December 1965 (AD 479–783).
5. Matyja, H., Giessen, B.C. and Grant, N.J., "The Effect of Cooling Rate on the Dendrite Spacing in Splat-Cooled Aluminum Alloys," *J. Inst. Metals*, 96 (1968), 30.
6. Sprowls, D.O., Walsh, J.D. and Shumaker, M.B., "Simplified Exfoliation Testing of Aluminum Alloys," Symposium on Localized Corrosion, ASTM Annual Meeting, Atlantic City, New Jersey, June 1971.
7. Fink, W.L. and Willey, L.A., "Quenching of 755 Aluminum Alloy," *Trans. AIME*, 175 (1948), 414.
8. Cebulak, W.S. and Truax, D.J., "Program to Develop High-Strength Aluminum Powder Metallurgy Products: Phase II–P/M Alloy Optimization," Alcoa Research Laboratories, New Kensington, Pennsylvania, Frankford Arsenal Contract Report, June 1971 (AD 884–642L).

PROCESSING OF WROUGHT PRODUCTS (B)

MODERATOR: FRANK ZALESKI
Frankford Arsenal
Philadelphia, Pennsylvania

13. Cold Rolling of Dispersion-Strengthened Nickel

L. F. NORRIS, R. W. FRASER, and D. J. I. EVANS
Sherritt Gordon Mines Limited
Fort Saskatchewan, Alberta

ABSTRACT

The cold rolling of dispersion-strengthened nickel is considered in terms of the mechanical aspects of rolling and the metallurgical response to rolling. Metallurgical response in this alloy is measured by the development of microstructures which can be characterized by the statistical parameters of the dispersoid distribution and by the grain size, grain shape, and grain orientation. The processing of a series of nickel-thoria powders to strip is used to illustrate the functional dependence of the physical and mechanical properties on the microstructure.

Introduction

Dispersion-strenthened nickel sheet is currently specified for certain critical components in advanced aircraft turbine engines because it offers significant advantages in strength, creep resistance, and fatigue life over competing superalloys at service termperature of 2000°F (1095°C) and higher [1]. Of particular importance in these and other candidate applications for dispersion-strengthened nickel and nickel-base alloys is the remarkable microstructural stability of these materials as manifested by the absence of solid-state solution reactions and the resistance to changes in grain size and grain structure. The key to this behavior is the presence in the chemically simple matrix of a fine, well-distributed dispersoid which stabilizes the microstructure. The special role of cold rolling in the production of dispersion-strengthened nickel sheet is that it is in this step of the processing that the characteristic grain structure is determined. The magnitude of the reduction and the rolling direction combine with the influence of the dispersoid distribution to establish specific microstructures, each with a unique combination of mechanical properties. The relationships between the mechanical and metallurgical aspects of cold rolling and the microstructure of the finished sheet or strip are the subject of this chapter.

The dual elements of microstructure are the characteristics of the dispersoid and the grain structure. In the absence of a single, simple quantitative expression for characterizing the dispersoid, the parameters of most interest are its weight or volume fraction, the distribution of particle sizes (including the mean, median, and mode sizes), and the spatial distribution of the dispersoid particles of which the interparticle spacing is one measure. The grain structure includes measures of grain size, grain shape, and grain orientation (texture). These elements of microstructure can be monitored to illustrate the relationships between the processing and the product mentioned above.

Relatively little of the extensive literature on dispersion-strengthened metals and alloys has dealt with rolling to sheet, reporting rather on extrusion or forging for primary working, and drawing or swaging for secondary working. Mee and Sinclair [2] reviewed previously published analyses of the texture of cold rolled nickel-thoria [3,4] and studied the effect of thoria particle dispersions on texture development during the cold rolling and annealing of extruded and extruded and cold-forged stock. All of the cold rolling in this study was unidirectional and to high reductions (65-98.6 percent) resulting in a pure-metal type of texture $\{135\}\langle 112\rangle$ which transformed on annealing to a variety of textures including $\{100\}\langle 001\rangle$, $\{210\}\langle 112\rangle$, $\{359\}\langle 301\rangle$, and $\{445\}\langle 1247\rangle$, depending on the nature of the thoria distribution and the extent of deformation.

Anders [5] noted the development of an oriented crystal structure in nickel-thoria sheet which has been cold rolled 40–95 percent and annealed above its recrystallization temperature. This observation was discussed more explicitly by his colleagues Baldwin et al. [6]. The latter hot worked nickel-thoria billets by extrusion or forging and then cold rolled the intermediate product a minimum of 30 percent and recrystallized. Sheets which had been consolidated by extrusion and cold rolled at least 30 percent developed a reported $\{321\}\langle 121\rangle$ crystallographic orientation (erroneous description, should be $\{321\}\langle 412\rangle$) of high purity, while sheets processed by hot forging and then rolled to a minimum of 85 percent reduction developed a similar pure $\{100\}\langle 001\rangle$ texture. Other textures reported were $\{110\}\langle 112\rangle$ and $\{110\}\langle 001\rangle$.

Fraser et al. [7,8] showed that the strength properties of nickel-thoria strip could be enhanced by cyclic cold rolling, each reduction of 10–20 percent per cycle being followed by annealing at 2200°F (1200°C). The ultimate tensile strength at 1600°F (870°C) increased from 17,000 psi (as hot consolidated by rolling) to 26,000 psi after 20 cold roll/anneal cycles. This rolling schedule resulted in a fine fibrous microstructure from which a strengthening model based on grain-boundary transfer of an applied load between fibers strengthened by the dispersoid was developed [9].

Wilcox et al. [10] recently correlated high-temperature strength properties (yield strength, stress to rupture, and creep strength) with grain-aspect ratio for Ni–ThO_2, Ni–Cr–ThO_2, Ni–Cr–W–ThO_2, and dispersion-strengthened superalloys.

The linear increase in these properties at 2000°F (1095°C) with increasing aspect ratio was observed to hold both for coarse-grain recrystallized material and fine-grain unrecrystallized material, and to provide a plausible complement to the grain-size effect historically considered the dominant criterion in high-temperature strength properties.

The present study was undertaken to explore the relationships between texture, grain structure, and short-time mechanical properties established in cold rolling nickel strip or sheet containing about 2 percent thoria by weight.

Materials and Procedures

Powder Preparation and Characterization

A number of hydrometallurgical processes have been developed at Sherritt Gordon for the preparation of nickel-thoria powders. One of these consists of using hydrogen at elevated temperatures and pressures to reduce an aqueous slurry of basic nickel carbonate in the presence of a sized thoria sol [7,11]. In another process [12], thorium nitrate is added to the basic nickel carbonate, forming a thorium-containing precipitate, which is essentially thorium hydroxide and nickel nitrate. The aqueous slurry of basic nickel carbonate particles with the adsorbed thorium compound is treated with hydrogen at elevated temperatures and pressures to effect reduction to elemental nickel. The reduced product is dried and then further treated in hydrogen in the temperature range 1500–1800°F (820-980°C) to convert the thorium compound to thoria. Proprietary modifications of these processes were used to prepare the nickel-thoria powders for this study.

All powders contained 2.0 or 2.1 percent thoria (by weight) and had the following analysis:

Carbon	0.005 percent
Sulphur	0.001 percent
Iron	0.02 percent
Other impurities	0.01 percent
Thoria	2.0 or 2.1 percent
Nickel	Balance

The thoria particle-size distributions were determined in the annealed strip product by sizing about 2,000 particles in transmission electron micrographs at 48,000 diameters magnification, such as shown in Figure 1. A Zeiss Particle Size Analyser TGZ-3 was used to count and size the particles. This method provided accurate data for particle sizes greater than about 0.009 micron. Omission of particles smaller than 0.009 micron did not affect the consistency of the

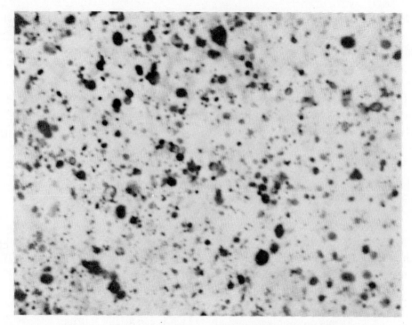

Figure 1. Thoria distribution in Ni–2.1% ThO$_2$ strip. 48,000X.

analyses. Typical dispersoid parameters for samples in this study, taken from a representative cumulative size distribution (Figure 2), were as follows:

Mean size   0.020 micron
Median size   0.013 micron
Mode size   0.011 micron
Interparticle spacing   0.09 micron

$$\text{IPS} = d \left[\left(\frac{\pi}{6f} \right)^{1/2} - \frac{\pi}{4} \right]$$

where   [13]

d = mean dispersoid diameter
f = volume fraction

Preparation and Characterization of Plate for Cold Rolling Experiments

Consolidation of the powders and primary working consisted of pressing into rectangular billets, sintering, and hot rolling. Small billets (2.4″ × 1.25″ × 0.2″) were prepared by static compaction at 33 tsi in a double-acting die; larger billets

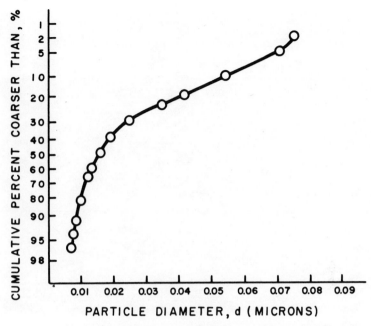

Figure 2. Cumulative ThO_2 particle-size distribution in Ni–2.1% ThO_2 strip.

(6″ × 3″ × 0.5″) were isostatically pressed at 15 tsi. The billets were sintered for two hours at 2000°F (1095°C) and then hot rolled with preheating in ultra-pure hydrogen at 2200°F (1200°C) for each pass. Hot rolling was done on a Loma two-high rolling mill with 8-inch-diameter rolls.

The starting material for the cold rolling experiments was usually stress-relieved for 30 minutes at 2000°F (1095°C). No recrystallization occurred during this anneal. As no grain structure could be resolved by optical micro-scopy, a collodion replica of a polished and etched section was prepared for examination in an electron microscope. The resulting photomicrograph (Figure 3) indicated that the grain size was of the order of 1-2 microns. As hot rolled, the plate exhibited a preferred crystallographic orientation. The strongest texture was {110} ⟨112⟩, a cold worked structure, indicating that some of the energy of deformation was retained in the material. Metallurgically, therefore, the rolling preheat temperature must have been below the recrystallization tem-perature during the primary working.

Cold Rolling Experiments

Cold rolling was done without lubrication on a Stanat two-high mill with rolls 4 inches in diameter or with lubrication on a Loma two-high mill with 8 inch

Figure 3. Electron micrograph (replica) of hot rolled nickel-thoria strip. 12,000X.

rolls, as will be noted for each series of experiments. With the Loma mill, kerosene was swabbed onto the rolls and the strip before each pass. In all cases, the strip was rolled about 2-3 percent per pass.

Ni–2.1% ThO_2 plate which had been hot rolled from an isostatically compacted billet and pure nickel strip (Sherritt Gordon's SG 100, analyzing 99.9 percent nickel, 0.01 percent carbon, 0.0004 percent sulphur) were cold rolled from 10-90 percent to obtain a measure of their work-hardening rates.

A series of statically compacted and sintered billets was hot rolled to a 70-percent reduction in thickness and then cold rolled on the Loma mill to trace the texture development in Ni–2.0% ThO_2 strip. Cold rolling reductions of 40, 50, 60, 70, and 80 percent parallel or perpendicular to the hot rolling direction were used. The texture components were analyzed both before and after recrystallization by annealing at 2500°F (1370°C).

The thoria size distributions of two experimental Ni–2.1% ThO_2 powders were tailored to illustrate the effect of the dispersoid distribution on the microstructure of the wrought product. The mean particle sizes of these powders, designated A and B, were 0.034 micron and 0.028 micron, respectively. The higher mean particle size of powder A indicated that the coarse particles in the distribution were larger than their counterparts in powder B. Both powders were fabricated to strip from a statically compacted billet by a 70-percent hot reduction followed by an 80-percent cold reduction on the Loma mill, either parallel to or perpendicular to the hot rolling direction. The texture components in the as-cold rolled strip and in samples which had been recrystallized by annealing at 2500°F (1370°C) were analyzed by X-ray diffraction.

Samples prepared to show the effect of increasing cold reduction on the microstructure and the mechanical properties of Ni–2.1% ThO_2 strip were processed from isostatically compacted billets hot rolled to a total reduction in thickness of 82 percent. Following a stress-relieving anneal, the plate was cold rolled 20, 40, 60, or 80 percent on the Stanat mill in a direction parallel or perpendicular to the hot rolling direction. The resulting strip was recrystallized by annealing at 2450°F (1345°C).

Analytical Procedures

Tensile testing was done on a Tinius Olsen tester rated at 10,000 lbs maximum load. Test specimens had gauge dimensions of $1'' \times 0.25''$, and were pulled at a constant crosshead speed of 0.02 inch per minute to failure. The 0.2 percent offset yield strength, the ultimate tensile strength, and the elongation in 1 inch were recorded in each test. Duplicate tests established an over-all precision of approximately ±2 percent in the strength values, reflecting both the precision of the test measurements and the sample-to-sample variations.

The designations "longitudinal" and "transverse" refer to the orientation of the samples with respect to the cold rolling direction.

Most texture data were obtained from the normalized intensities of the {111}, {100}, {110}, {113}, {120}, and {112}, reflections by X-ray diffractometry. The computer-calculated data were rounded to the nearest percent for presentation. For these results, the first order of {hkl} was used to represent the family whether or not the first-order reflection is allowed by the structure factor. Selected textures were confirmed by preparation of {110} pole figures using a Philips Texture Goniometer on a Philips 1010 X-ray generator with a copper X-ray tube.

Standard metallographic mounting, polishing, and etching procedures were used in the preparation of optical micrographs. Specimens were etched in two steps. The first etchant, a mixture of 10 parts nitric acid, 5 parts acetic acid and 85 parts water (by volume), was applied electrolytically at 1.5 volts for one minute. This was followed by immersion for 5 to 10 seconds in a mixture of equal volumes of nitric acid and acetic acid.

Experimental Results

Mechanical Aspects of Rolling

The work-hardening characteristics of the dispersion-strengthened nickel plate are compared with those for pure nickel in Figures 4 and 5. The 0.2-percent

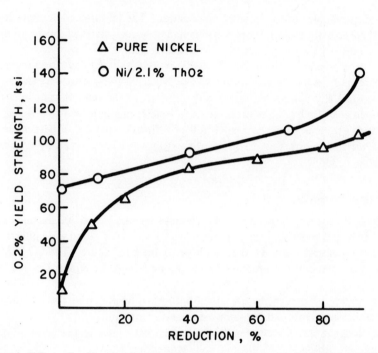

Figure 4. Effect of cold reduction on yield strength of pure nickel and Ni–2.1% ThO$_2$ at room temperature.

yield strength of the dispersion-strengthened nickel increased more slowly at cold reductions below 20 percent but then it maintained a similar work-hardening rate in the range from 20 percent to about 70 percent. The initial ductility of the dispersion-strengthened nickel plate was only about one-half that of the pure nickel plate, but a greater percentage of the ductility was retained with reductions to 90 percent. Thus, the cold rolling of Ni–2% ThO$_2$ is similar to the cold rolling of pure nickel. The roll-separating forces demanded in rolling the former are only slightly higher than those for pure nickel, and are considerably lower than those for common nickel-base super-alloys. Similarly, the ductility imposes no operational difficulty and edge cracking, for example, is negligible with reductions in excess of 80 percent.

Effect of Cold Rolling on Texture Development

The as-rolled texture composition of samples cold rolled parallel to or perpendicular to the hot rolling direction is summarized in Table I. Compared with the starting material, the percentage of the {111}, {100}, and {120} components

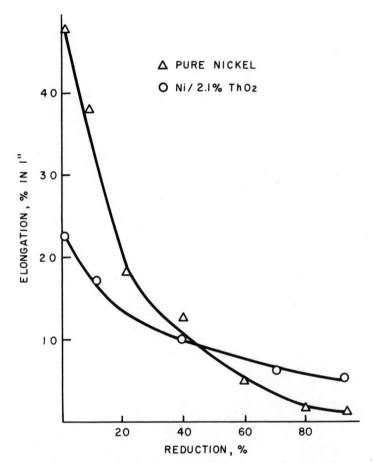

Figure 5. Effect of cold reduction on tensile elongation of pure nickel and Ni–2.1% ThO$_2$ at room temperature.

decreased and that of the {110} component increased. However, there were few changes in texture composition within the 40-80 percent range of reductions, irrespective of rolling direction.   For both rolling directions, the amount of {112} component increased and the {120} decreased with increasing reduction; these two components accounted for less than 20 percent of the total, however.

A recrystallization anneal of these samples resulted in the textures given in Table II. All strip cold rolled parallel to the hot rolling direction had a strong {120} orientation with only the sample rolled 80 percent showing an appreciable amount of the cube texture. Increased amounts of cross-rolling, on the other hand, led to development of a mixed texture comprising principally {100} and {112}.

TABLE I

Distribution of Texture Components in As-Cold Rolled Ni–2% ThO_2 Strip

Cold Reduction (%)	Normalized Intensities from Diffractometer						
	{111} (%)	{100} (%)	{110} (%)	{113} (%)	{133} (%)	{120} (%)	{112} (%)
Cold Rolled Parallel to Hot Rolling Direction							
0	3	10	47	17	7	12	4
40	0	4	68	20	1	5	2
50	0	4	65	21	1	5	4
60	0	2	70	18	1	3	6
70	0	2	67	18	1	3	9
80	1	3	65	13	0	2	16
Cold Rolled Perpendicular to Hot Rolling Direction							
0	3	10	47	17	7	12	4
40	1	5	56	15	6	14	3
50	0	4	61	18	4	10	3
60	0	3	56	22	5	8	6
70	0	1	62	71	3	5	8
80	0	0	61	20	3	3	13

TABLE II

Distribution of Texture Components in Cold Rolled and Recrystallized Ni–2% ThO_2 Strip

Cold Reduction (%)	Normalized Intensities from Diffractometer						
	{111} (%)	{100} (%)	{110} (%)	{113} (%)	{133} (%)	{120} (%)	{112} (%)
Cold Rolled Parallel to Hot Rolling Direction							
0	2	9	51	17	5	12	4
40	1	4	0	0	2	91	2
50	0	4	0	0	3	91	2
60	0	3	0	0	3	92	2
70	0	3	0	0	2	95	0
80	0	13	0	0	2	85	0
Cold Rolled Perpendicular to Hot Rolling Direction							
0	2	9	51	17	5	12	4
40	0	2	80	4	7	7	0
50	0	10	60	5	10	10	5
60	0	30	15	3	12	20	20
70	0	85	0	0	3	2	10
80	0	82	6	0	1	0	11

The texture composition of strip fabricated from powders A and B and cold rolled 80 percent is given in Table III. Changes in the thoria size distribution had no apparent effect in the case of uniaxial rolling, further emphasizing the insensitivity of the uniaxial cold rolling process. The cross-rolled samples had similar texture compositions, except for the {110} component which increased with a finer thoria distribution.

TABLE III

Distribution of Texture Components in As-Cold Rolled Ni–2% ThO$_2$ Strip
Fabricated from Tailored Powders A and B

Powder	Thoria Size		Normalized Intensities from Diffractometer						
	Mean (micron)	Median (micron)	{111} (%)	{100} (%)	{110} (%)	{113} (%)	{133} (%)	{120} (%)	{112} (%)
			Cold Rolled Parallel to Hot Rolling Direction						
A	0.034	0.016	0	4	72	11	0	1	12
B	0.023	0.014	0	1	76	11	0	2	10
			Cold Rolled Perpendicular to Hot Rolling Direction						
A	0.034	0.016	0	9	44	20	3	6	18
B	0.023	0.014	0	3	65	17	0	2	13

The recrystallized samples showed substantial differences in texture composition (Table IV). Strong cube textures were formed in the strip from the coarser powder A rolled uniaxially and in the strip with the finer thoria (powder B) which had been cross-rolled. Uniaxial rolling of material from powder B resulted in a mixture of the {120} ⟨001⟩ and {112} ⟨312⟩ components. Other texture

TABLE IV

Distribution of Texture Components in Cold Rolled and Recrystallized Ni-2.1% ThO$_2$ Strip
Fabricated from Tailored Powders A and B

Powder	Thoria Size		Normalized Intensities from Diffractometer						
	Mean (microns)	Median (microns)	{111} (%)	{100} (%)	{110} (%)	{113} (%)	{133} (%)	{120} (%)	{112} (%)
			Cold Rolled Parallel to Hot Rolling Direction						
A	0.034	0.016	0	96	0	0	3	1	0
B	0.023	0.014	0	2	0	0	2	62	34
			Cold Rolled Perpendicular to Hot Rolling Direction						
A	0.034	0.016	1	80	3	2	7	2	5
B	0.023	0.014	0	93	0	0	1	0	6

components appeared during annealing of the cross-rolled samples from A, although the cube texture was dominant.

Effect of Cold Rolling on Grain Size and Shape

The characteristic grain structures developed in the strip cold rolled 20-80 percent and recrystallized are illustrated in Figures 6 and 7. For samples rolled uniaxially, increasing reductions led to pronounced fibering of the grains along the rolling direction (Figure 6). Since the number of grains in the thickness direction simultaneously increased, the increase in aspect ratio was associated with a decrease in grain size as well. Grains sectioned transversely to the rolling direction had a lower and relatively constant aspect ratio.

Larger differences in grain size and shape were observed in the microstructures of the cross-rolled samples. Strip rolled 20, 40, 60, or 70 percent developed a duplex microstructure on recrystallization, with bands of coarser grains at the surface (Figure 7). The aspect ratios of these outer grains were similar in the longitudinal and transverse sections and were comparable to those in the transverse sections of the uniaxially rolled material. Grains in the central structure were also elongated, and varied in size with the magnitude of the cold reduction. The grain size was highest for the 20-percent reduction, decreased substantially after reductions of 40 or 60 percent, and then increased with the highest reduction. As with the unidirectionally rolled samples, the microstructures in the longitudinal and transverse sections were similar.

Pole figure analysis confirmed the presence in these samples of the texture components identified in the previous section for the various cold reductions.

Effect of Cold Rolling on Mechanical Properties

The effects of cold rolling on the tensile strength and ductility of Ni-2.1% ThO_2 strip are summarized in Tables V and VI and Figures 8, 9, and 10. At room temperature, the 0.2-percent yield strengths of the longitudinal samples gradually decreased with increasing reduction, whereas the transverse sample exhibited a peak between 20-percent and 60-percent reduction. The anisotropy of the uniaxially rolled material reversed between the 40-percent and 60-percent reduction; there was no consistency in the anisotropy of the cross-rolled samples. The corresponding ultimate tensile strengths generally showed a maximum, with the sharpest peaks occurring in the stronger cross-rolled series. Again there was no pattern in the anisotropy of the cross-rolled sample but, for the uniaxially rolled material, the difference by which strength in the longitudinal direction exceeded that in the transverse direction increased from 1,000 psi at 20-percent

Longitudinal Section   Transverse Section

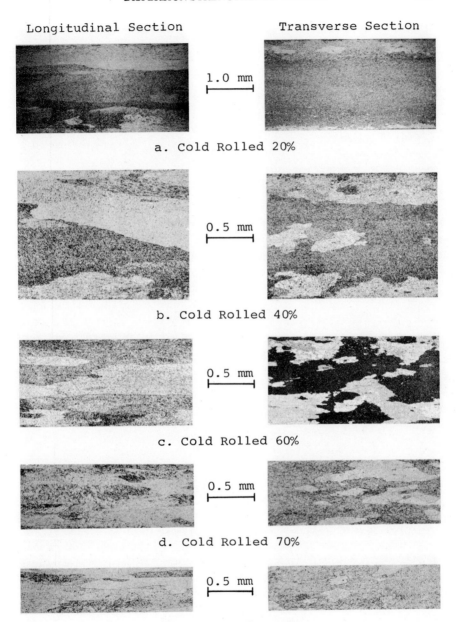

1.0 mm

a. Cold Rolled 20%

0.5 mm

b. Cold Rolled 40%

0.5 mm

c. Cold Rolled 60%

0.5 mm

d. Cold Rolled 70%

0.5 mm

e. Cold Rolled 80%

Figure 6. Micrographs of Ni–2.1% ThO$_2$ strip cold rolled parallel to hot rolling direction and recrystallized.

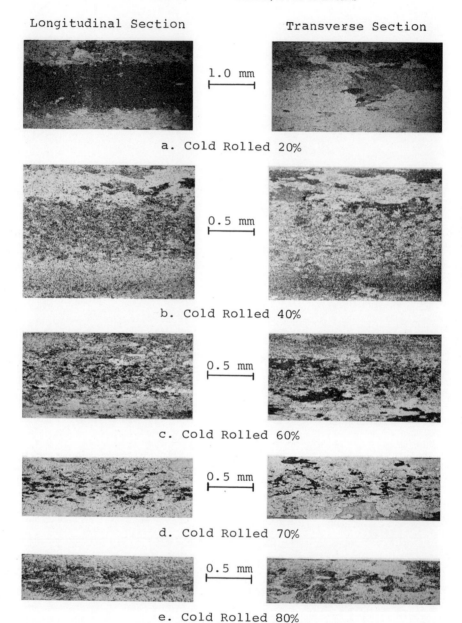

Figure 7. Micrographs of Ni–2.1% ThO$_2$ strip cold rolled perpendicular to hot rolling direction and recrystallized.

TABLE V

Room-Temperature Tensile Test Data for Ni–2.1% ThO$_2$ Strip
Cold Rolled 20–80% and Recrystallized

Cold Reduction (%)	Test Direction	0.2% Yield Strength (psi)	Ultimate Strength (psi)	Elongation (% in 1″)
	Cold Rolled Parallel to Hot Rolling Direction			
20	Longitudinal	46,600	71,000	17
	Transverse	48,000	69,900	25
40	Longitudinal	45,900	73,400	17
	Transverse	49,400	72,200	23
60	Longitudinal	45,300	73,400	17
	Transverse	44,200	70,700	23
70	Longitudinal	44,500	72,700	15
	Transverse	41,900	68,900	23
80	Longitudinal	43,200	72,000	14
	Transverse	40,900	67,100	19
	Cold Rolled Perpendicular to Hot Rolling Direction			
20	Longitudinal	51,200	76,200	27
	Transverse	45,000	72,900	20
40	Longitudinal	51,600	78,500	20
	Transverse	53,900	78,600	25
60	Longitudinal	49,500	79,200	21
	Transverse	50,600	80,300	25
70	Longitudinal	49,000	80,200	22
	Transverse	46,400	77,900	21
80	Longitudinal	45,200	76,500	16
	Transverse	45,700	75,800	18

reduction to nearly 5,000 psi at 80-percent reduction. The room-temperature ductilities (Figure 9), as measured by tensile elongation, ranged from 14 percent to 27 percent within trends resembling those in the 0.2-yield-strength data. The ductilities in the uniaxially rolled material decreased with relatively constant anisotropy as the reduction increased. For the cross-rolled samples, both the ductility and the anisotropy in ductility generally decreased.

At 2000°F (1095°C) the 0.2-percent yield strength (Figure 10) and the ultimate tensile strength of samples which had been cross-rolled exhibited pronounced minima at a reduction between 40 and 70 percent. The uniaxially rolled material showed more gradual changes in strength and a reversal in anisotropy with increasing reduction. The ductility was virtually constant over the entire range.

TABLE VI

2000° F Tensile Test Data for Ni–2.1% ThO$_2$ Strip
Cold Rolled 20–80% and Recrystallized

Cold Reduction (%)	Test direction	0.2% Yield Strength (psi)	Ultimate Strength (psi)	Elongation (% in 1″)
	Cold Rolled Parallel to Hot Rolling Direction			
20	Longitudinal	16,400	18,300	3
	Transverse	18,500	19,000	2
40	Longitudinal	16,700	19,700	2
	Transverse	17,900	18,600	2
60	Longitudinal	17,200	18,700	2
	Transverse	17,100	18,500	2
70	Longitudinal	17,400	18,600	2
	Transverse	17,300	17,600	2
80	Longitudinal	17,100	18,300	2
	Transverse	16,400	16,400	2
	Cold Rolled Perpendicular to Hot Rolling Direction			
20	Longitudinal	18,000	18,800	2
	Transverse	16,800	18,200	3
40	Longitudinal	14,700	16,600	2
	Transverse	15,700	17,900	3
60	Longitudinal	14,200	15,500	2
	Transverse	14,900	16,000	2
70	Longitudinal	17,400	18,000	2
	Transverse	16,300	17,500	2
80	Longitudinal	18,000	18,500	2
	Transverse	17,100	17,300	2

Discussion

Dispersion-strengthened nickel strip processed by cold rolling 20-80 percent and recrystallizing develops specific texture compositions unlike pure nickel strip similarly processed. The relationships between preferred crystallographic orientation, grain structure, and short-time mechanical properties of Ni–ThO$_2$ strip at 2000°F (1095°C), each of which was measured as a function of cold rolling, can be illustrated by comparing rankings of grain size (based on grain-boundary length) and grain aspect ratio (grain length/grain width) and yield strength and then relating the degree of consistency observed to the character-istics of the textures measured. The cases for the two cold rolling directions are considered separately.

The rankings for samples for which the cold rolling direction was parallel to the hot rolling direction are given in Table VII. Averages of the values for the longitudinal and transverse directions for each entry have been used for this

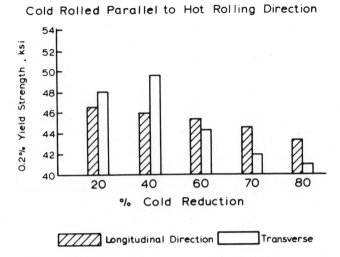

Cold Rolled Parallel to Hot Rolling Direction

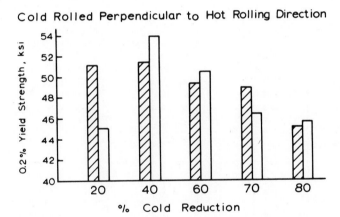

Cold Rolled Perpendicular to Hot Rolling Direction

Figure 8. Yield strength at room temperature of Ni–2.1% ThO$_2$ strip cold rolled 20–80% and recrystallized.

comparison, notwithstanding the bias from the more fibrous microstructure in the longitudinal section of the highly rolled samples. The three rankings indicate a correlation between increasing high-temperature yield strength and increasing average grain aspect ratio; the grain size was of less significance in these microstructures. The general trends in yield-strength level in each test direction for each reduction and in anisotropy appeared to correlate with development of an oriented {120} ⟨001⟩ texture and, to a lesser extent, with the increase in grain aspect ratio. Increasing strength and purity of the {120} ⟨001⟩ texture has been found to be associated with increasing high-temperature yield strength

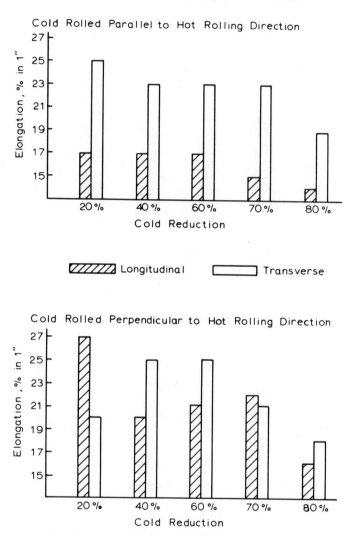

Figure 9. Tensile elongation at room temperature of Ni–2.1% ThO$_2$ strip cold rolled 20–80% and recrystallized.

along the rolling direction and with decreasing strength in the transverse direction. Rigorous interpretation of these strength/texture relationships must await further experimental work in which the role of the minor texture components can be assessed.

The texture development and the grain aspect ratio together determine the changes in strength anisotropy, but their relative contributions could only be inferred. The appearance of isotropic components such as {100} ⟨001⟩ in the

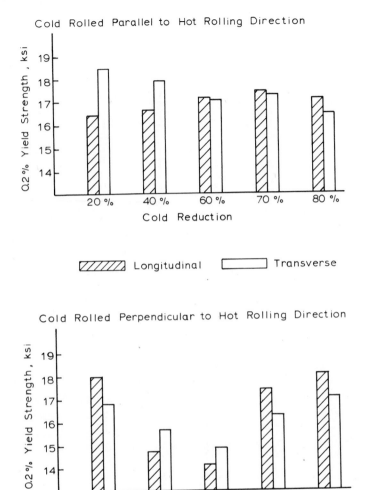

Figure 10. Yield strengths at 2000°F of Ni–2.1% ThO$_2$ strip cold rolled 20–80% and recrystallized.

texture composition at high reductions suggested that the aspect ratio was of increasing importance in this range.

The ranking of averaged data for grain size, grain aspect ratio, and yield strength for material processed by cross-cold rolling is given in Table VIII. Here the changes in 0.2-percent yield strength correlate better with average grain size, the lowest strength being associated with the reductions yielding the smallest grain sizes. The lack of consistency between the correlations in Table VII (grain

TABLE VII

Ranking of Average Grain Size, Grain Aspect Ratio, and Yield
Strength at 2000° F (1095° C) for Uniaxially Cold Rolled
Ni–2.1% ThO$_2$ Strip

	Ranking by Percent Cold Reduction		
	Average Grain Size	Average Grain-Aspect Ratio	Average 0.2% Yield Strength at 2000° F
Increasing Value ↑	20	* ⎡20	20
	40	⎣70	* ⎡70
	60	40	⎣40
	70	* ⎡60	60
	80	⎣80	80

*Equal rank.

aspect ratio) and Table VIII (grain size) suggests that texture exerts a stronger influence on the strength of Ni–ThO$_2$ alloys at high temperature than has previously been appreciated. The texture composition in this case was more complex than for uniaxial rolling, and a correlation with texture was further complicated by large differences in texture strength. The lowest values of yield strength (see Figure 10) occurred in a range of reductions that was associated with several low-order texture components, all of them relatively weak. (Strong texture components are considered to be those whose strength is 25-30 times that in a randomly oriented microstructure; weak textures have strengths only 3-5 times random.) The increase in strength at the higher reductions thus appeared to be due to development of a well-defined texture. The presence of {120} ⟨001⟩ in the surface layers may have been responsible for the anisotropy which developed

TABLE VIII

Ranking of Average Grain Size, Grain Aspect Ratio, and Yield
Strength at 2000° F (1095° C) for Cross-Cold Rolled
Ni–2.1% ThO$_2$ Strip

	Ranking by Percent Cold Reduction		
	Average Grain Size	Average Grain-Aspect Ratio	Average 0.2% Yield Strength at 2000° F
Increasing Value ↑	20	* ⎡80	* ⎡80
	80	⎣70	⎣20
	70	60	70
	60	* ⎡20	* ⎡40
	40	⎣40	⎣60

*Equal rank.

simultaneously. The specific contributions of the $\{112\}$ component and various higher order components such as $\{130\} \langle 001 \rangle$ which have been identified in cross-rolled samples (only), have not been determined.

Changes in the dispersoid distribution radically change the texture composition, as was demonstrated with tailored powders A and B (Table IV). Thus, while the magnitude of the cold reduction and its direction determine the texture and grain morphology, the dispersoid distribution appears to regulate the response to the cold rolling, as seen in the recrystallized microstructure. Definition of the relationships between the cold rolling parameters and the grain structure must therefore also include characterization of the thoria distribution. An analytical approach to this definition will require further studies of the characterization of the dispersoid distribution and the development of textured microstructures.

The textures developed during cold rolling in the present study by specification of the rolling direction and/or the thoria distribution arise from different recrystallization mechanisms. For the case of cross-rolling, it is postulated that high deformation leads to the formation of regions of high stress gradient and considerable misorientation. These provide the driving force for growth of nuclei by high-angle grain-boundary migration into the cold worked matrix, resulting in a recrystallization texture of totally different orientation from the majority of the original cold worked matrix. The decrease in cube component in the as-rolled texture with reduction (Table I) and its increase in the annealed structure suggest a recrystallization model rather than grain coalescence.

Low deformation apparently does not produce these nuclei, and a recovery and growth process takes place instead. The cold worked structure polygonizes to form a subgrain structure. Growth takes place by subgrain coalescence, giving a final plane orientation identical to the original matrix orientation. Tables I and II show that the $\{110\}$ orientation is favored by this process. In the cold worked matrix, $\{110\}$ and $\{113\}$ regions exist in similar proportions, but in the annealed matrix $\{110\}$ dominates. This low deformation process only applies to the cross-rolled fabrication; uniaxial fabrications produce a $\{120\}$ annealed texture regardless of the amount of cold reduction.

With uniaxial rolling, a more perfectly oriented subgrain structure would be expected, and driving forces for grain growth due to boundary misfits would consequently be less. Since the $\{120\} \langle 001 \rangle$ annealed texture is related to the parent rolling texture $\{110\} \langle 112 \rangle$ by $45°$ rotations about $\langle 111 \rangle$ directions, atom transfer across a boundary or through a series of intermediate orientations may be, in this case, the most energetically favorable recrystallization mechanism.

The principal texture components for the two cold rolling schemes previously discussed, $\{100\}$ and $\{120\}$, have been considered to be primary recrystallization textures. Each developed directly from the cold-worked matrix and reached its full strength (for the given prior processing) within two minutes. Subsequent annealing for up to 16 hours at 2500°F (1370°C) did not alter the texture strengths. Stability of the texture implies over-all stability of the microstructure and this, too, has been demonstrated. The tensile properties at room temperature and at 2000°F (1095°C) were unchanged after annealing at 2500°F (1370°C) for periods up to 106 hours.

Properties other than tensile strength are also of interest in engineering applications of dispersion-strengthened nickel strip or sheet. For example, formability at room temperature as measured by tensile elongation, cup ductility, or bend ductility is determined by the textural composition and the grain size. Higher tensile elongations and cup heights have been found to be associated with material having a strong {120} texture. However, this strip had poor bend ductility and its generally coarser grain structure gave rise to surface roughening (orange peel) during deformation. Cross-cold reduction in processing yielded a cube texture and provided the desirable finer grain structure with acceptable values for elongation and cup ductility.

The extent of anisotropy in ductility is determined by the isotropy of the texture components in the matrix. Irrespective of the changing grain aspect ratio, there was considerable anisotropy in tensile elongation in the samples which were cold rolled in the hot rolling direction (see Table V). The cross-rolled material, on the other hand, exhibited decreasing anisotropy coincident with the development of a {100} texture.

Even stronger effects from the texture composition and strength have been observed in studies of variations in the elastic modulus. Values from 18×10^6 psi to over 32×10^6 psi have been recorded in experimental material having a fixed thoria content (2.1 percent) and yield strength (at 2000°F) of 14,000–17,000 psi. In general, the lowest average moduli and most isotropic moduli have been measured in microstructures with the strongest cube texture. However, strip with a very strong {120} ⟨001⟩ texture has a low modulus (18×10^6–22×10^6 psi) in the longitudinal direction but a high modulus in the transverse direction (28×10^6–32×10^6 psi). These values are consistent with the observation that a decrease in modulus is associated with stronger ⟨001⟩ texture components in the rolling direction.

Other structure-sensitive properties such as thermal and electrical conductivity, oxidation and hot corrosion resistance, magnetization and coercivity, and thermionic emission of dispersion-strengthened nickel can also be tailored, in principle, by control of the microstructure. Selection of the optimum dispersoid distribution in the starting powder and the processing route to the wrought product to achieve the desired combination of properties must be made empirically at present. Thus, a primary goal in the continuing research on dispersion-modified metals and alloys should be an understanding of the functional dependence of properties on the various aspects of microstructure. Further processing studies can then focus on development of specific microstructures.

Summary

The mechanical aspects of cold rolling Ni–ThO$_2$ strip have been shown to be similar to those of pure nickel, the only significant differences arising from the varying rates of work hardening. There are profound metallurgical changes during the cold rolling, however. Some of the relationships between texture development, grain size and shape, and the mechanical properties of cold rolled and recrystallized dispersion-strengthened nickel strip or sheet have been investigated. Several series of Ni–ThO$_2$ powders were fabricated to strip by hot rolling, cold rolling 20–80 percent, and annealing to recrystallize.

Strip which had been cold rolled parallel to the hot rolling direction: (1) developed a $\{120\}\langle 001\rangle$ texture, and texture strength increased with increasing cold reduction; (2) had strength characteristics at 2000°F (1095°C) which correlated with the grain aspect ratio; (3) exhibited increasing strength at 2000°F (1095°C) along the rolling direction; (4) had anisotropy in strength at 2000°F (1095°C) and in tensile elongation at room temperature; (5) had high and anisotropic room-temperature elastic moduli.

Strip for which the cold rolling direction was perpendicular to the hot rolling direction: (1) developed a $\{100\}\langle 001\rangle$ texture; (2) had strength characteristics at 2000°F (1095°C) which correlated with grain size; (3) exhibited minimum strength levels for intermediate rolling reductions of 40–60 percent, which yielded weakly textured microstructures; (4) had nearly isotropic strength at 2000°F (1095°C) and tensile elongation at room temperature; (5) had lower and isotropic room-temperature elastic moduli.

It was also shown that the texture composition was determined by the characteristics of the dispersoid distribution. Thus, a functional dependence of the mechanical properties of dispersion-strengthened nickel on various parameters of the microstructure such as the dispersoid distribution, the preferred crystallographic orientation, and the grain size and shape has been observed. Further understanding of these interrelationships will, in principle, permit tailoring of the properties of dispersion-modified metals and alloys by control of their microstructure.

Acknowledgments

The authors thank Mr. David D. Thomas, President, Sherritt Gordon Mines Limited, for permission to publish this paper. The work was carried out under a shared-costs research contract with the Defence Research Board of Canada under the Defence Industrial Research Programme. Due acknowledgement is given to the suggestions and assistance given by various members of the Physical Metallurgy Research Department, especially to Mr. M. J. H. Ruscoe who provided the texture analyses.

References

1. Bradley, E.F., Phinney, D.G. and Donachie, M.J. Jr., "The Pratt & Whitney Gas Turbine Story," *Metal Prog.*, 97 (1970), 68.
2. Mee, P.B. and Sinclair, R.A., "Cold-Rolling and Annealing Behavior in Nickel-Thoria Sheet, with Particular Reference to Preferred Orientation Development," *J. Inst. Metals*, 94 (1966), 319.
3. Inman, M.C., Zwilsky, K.M. and Boone, D.H., "Recrystallization Behavior of Cold Rolled TD-Nickel," *Trans. ASM*, 57 (1964), 701.
4. Clark, C.A. and Mee, P.B., "Verformungs und Rekristallisationstexturen von Nickel und von Legierungen auf Nickelbasis," *Z. Metallkunde*, 53 (1962), 756.
5. U.S. Patent No. 3,159,908, December 8, 1964, *Dispersioned Hardened Metal Product and Process*, Frederic J. Anders Jr. to E.I. du Pont de Nemours and Company, Wilmington, Delaware.
6. U.S. Patent No. 3,346,427, October 10, 1967, *Dispersioned Hardened Metal Sheet and Process*, William M. Baldwin Jr. *et al.*, to E.I. du Pont de Nemours and Company, Wilmington, Delaware.
7. Fraser, R.W., Meddings, B., Evans, D.J.I. and Mackiw, V.N., "Dispersion-Strengthened Nickel by Compaction and Rolling of Powder Produced by Pressure Hydrometallurgy," International Powder Metallurgy Conference, New York, 1965, in *Modern Developments in Powder Metallurgy*, Vol. 2, *Applications*, Henry H. Hausner, ed., New York: Plenum Press (1966), 87.
8. U.S. Patent No. 3,366,515, January 30, 1968, *Working Cycle for Dispersioned Materials*, Robert W. Fraser, and David J.I. Evans to Sherritt Gordon Mines Ltd., Toronto, Canada.
9. Fraser, R.W. and Evans, D.J.I., "The Strengthening Mechanism in Dispersion Strengthened Nickel," *Oxide Dispersion Strengthening*, G.S. Ansell, T.D. Cooper, and F.V. Lenel, eds., New York: Gordon and Breach, Science Publishers, Inc. (1968), 375.
10. Wilcox, B.A., Clauer, A.H. and Hutchinson, W.B., "Structural Stability and Mechanical Behavior of Thermochemically Processed Dispersioned Strengthened Nickel Alloys," Battelle Memorial Institute, Metal Science Group, Columbus, Ohio, NASA Contract Report CR–72832, March 1971 (N71–21589).
11. U.S. Patent No. 3,469,967, September 30, 1969, "Process for the Production of Nickel Refractory Oxide Powders and Production Thereof," Basil Meddings *et al.*, to Sherritt Gordon Mines Ltd., Toronto, Canada.
12. U.S. Patent No. 3,526,498, September 1, 1970, "Production of Nickel-Thoria Powders," David J.I. Evans and Bauke Weizenbach to Sherritt Gordon Mines Ltd., Toronto, Canada.
13. Durber, G.L.R. and Davies, T.J., "Strain Hardening in Polycrystalline Nickel Containing a Dispersion of Thoria," *Mat. Sci. Eng.*, 5 (1970), 142.

14. Development of IN-100 Powder-Metallurgy Disks for Advanced Jet Engine Application

R. L. ATHEY and J. B. MOORE
Pratt & Whitney Aircraft
West Palm Beach, Florida

ABSTRACT

Powder metallurgy is becoming increasingly important for highly alloyed nickel-base superalloys used for turbine and compressor disks in advanced jet engines. These alloys, by virtue of their alloying constituents, are very segregation-prone, particularly when cast into ingots large enough to produce engine disks; hence, the attractiveness of the powder-metallurgy route. Pratt & Whitney Aircraft has, through the development and utilization of an "all inert" powder-metallurgy concept and the *Gatorizing*^{TM} forging process, succeeded in producing turbine and compressor disks of IN-100 alloy for advanced engine use. This chapter reviews the powder-manufacturing processes used, the methods of powder consolidation applied, the interrelationship of the powder product and the *Gatorizing* forging process, and the attributes of the final disk product.

Introduction

This chapter is a review of Pratt & Whitney Aircraft progress in powder metallurgy for application of superalloys to jet engine disks.

During the early 1960s, in the development of certain advanced jet engines by Pratt & Whitney Aircraft, it became apparent that Waspaloy and René 41—the most advanced nickel-base alloys then used for forgings—did not possess the high-temperature strength required for the turbine disks. A new wrought alloy, at that time known as Astroloy, did exhibit the required strength and elevated-temperature operating capability. The evaluation of Astroloy forgings revealed that while the desired strength potential was present, these forgings exhibited excessive scatter in mechanical properties due to gross structural segregation and lack of homogeneity.

The segregation noted in early Astroloy forgings could be traced back to the large columnar grain structure of the ingots being produced at that time.

Metallurgical evidence in the form of macro- and microstructures of Astroloy, Waspaloy, and Inconel 718 from several major melting sources was compiled, showing that the segregation existing in conventional ingots was not eliminated by subsequent forging and thermal treatments. The influence of ingot structure on final disk microstructure and mechanical properties was more evident in Astroloy than in weaker alloys.

Based on these findings, a full-scale Astroloy ingot evaluation program was initiated by Pratt & Whitney Aircraft. The results of the evaluation showed that casting parameters yielding a uniformly fine, equiaxed grain structure minimized macrosegregation in the ingot. As a result of development aimed in this direction, several melting vendors are now producing ingots up to 12 inches in diameter with structure approaching that requested by Pratt & Whitney Aircraft. Disks from the improved ingot have substantially more uniform properties and microstructure.

Through the use of ultrafine equiaxed grain ingots (Figure 1), small disks were forged in IN-100 and MAR-M 200. The macro- and microstructure (Figure 2) of these disks, produced from what was considered an optimum fine-grain ingot, still contained a significant amount of segregation. The difficulty of producing this ultrafine-grain size in ingots large enough for aircraft engine disks is considered to be greater than that of producing powder billets.

The powder-metallurgy approach to the production of large engine disks represents an extension of the ingot evaluation program to produce material with a minimum of macrosegregation.

The degree of macrosegregation in a small powder particle (Figure 3) would be expected to be less by several orders of magnitude than that present in the structure of a large ingot conventionally melted and cast under optimum conditions. The amount of segregation in a large powder billet should be no more than that found in the individual powder particles making up the billet; therefore, in theory, there should be no limitation on powder-billet size with regard to segregation. We know that macrosegregation increases sharply with ingot diameter in conventionally melted and cast ingots, and eventually becomes prohibitive with superalloy compositions needed for advanced engines. The P/M approach is a most promising method for producing advanced wrought disks from the more highly segregation prone alloys such as Astroloy, IN-100, and AF2-1DA.

Initial Development Program

A method of powder production, collection, and densification was developed by Pratt & Whitney Aircraft which eliminated scatter in properties and lack of reproducibility, the problems associated with particle oxidation. To avoid oxida-

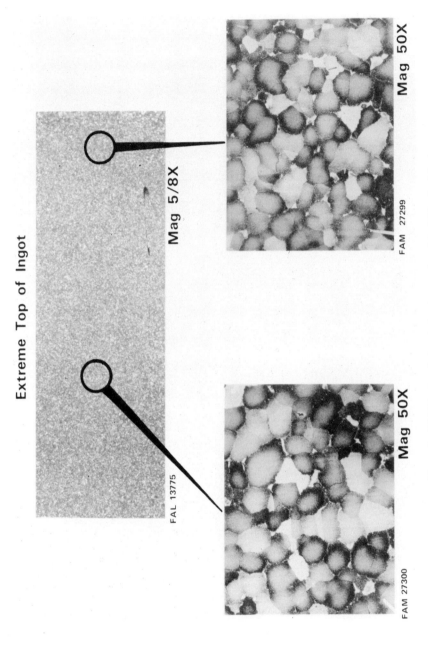

Figure 1. IN-100 subscale ingot structure, reduced to 85% of original.

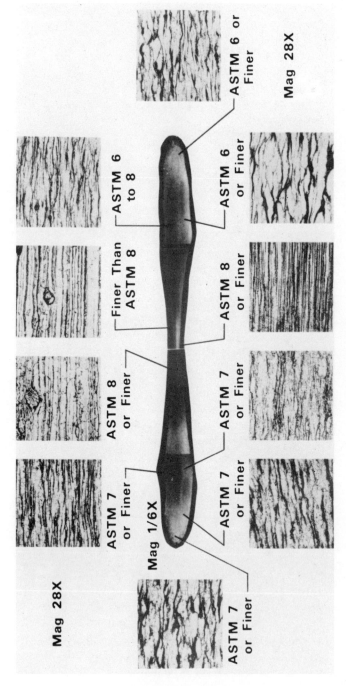

Figure 2. MAR-M 200 disk forged from fine-grain ingot, reduced to 75% of original.

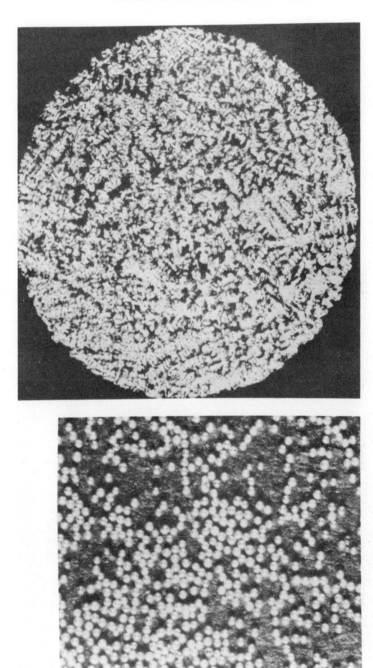

Mag 500X

Mag 40X

Figure 3. Spherical powder particles, 40X and 500X, respectively, reduced to 75% of original.

tion of individual powder particles, an "all inert" method of powder production, collection, and billet densification was developed wherein the metal powders would not be exposed to the atmosphere prior to densification. The investigation of this approach involved the production and evaluation of powder, small densified billets, and forgings produced from these billets. Astroloy, the most advanced production nickel-base superalloy disk material at that time, was selected for the investigation because of the excellent base line that Pratt & Whitney Aircraft had developed on conventially processed Astroloy disks from improved ingots and because the greatest gains using powder should be realized with such a highly alloyed material.

This initial program demonstrated that small Astroloy forgings made by the "all inert" P/M technique could be compacted and forged over a wide temperature range and showed uniformity of structure and mechanical properties. It was also found that oxygen contents well below 100 ppm could be consistently maintained.

The macro- and microstructural segregation of the small powder disks was significantly reduced from that of even the most optimum structure seen in a disk made from a cast ingot. The mechanical properties obtained from small Astroloy powder forgings are compared to PWA 1013 Astroloy disk-specification requirements (Figure 4). The strength and ductility, with minor deviations, exceeded the specification requirements.

TENSILE

Process	Temperature, °F	0.2% Yield Strength, ksi	Ultimate Strength, ksi	Elongation, %	Reduction Area, %
Powder Forging	RT	156.4	214.5	20.0	24.3
PWA 1013E	RT	140	195	16.0	18.0
Powder Forging	1400	140.4	161.7	18.7	38.1
PWA 1013E	1400	125	150	20.0	30.0

STRESS RUPTURE

Process	Temperature, °F	Stress, ksi	Life, hr	Elongation, %
Powder Forging	1400	85	66.1	21.1
PWA 1013E	1400	85	30.0	17.0

0.1% CREEP

Process	Temperature, °F	Stress, ksi	Life, hr	Elongation, %
Powder Forging	1300	74	200	–
PWA 1013E	1300	74	150	–
			Avg-110 Min	

Figure 4. Mechanical properties of small flat forging from Astroloy powder.

Scale-Up to Full-Size Engine Disks

The initial development effort on Astroloy was based on the direct forging of disks from compacted powder billets.

While this Astroloy powder program was underway, it was discovered at Pratt & Whitney Aircraft's Florida Research and Development Center that iron-, nickel-, and titanium-base superalloys could be placed in a condition of superplasticity.

Billet and bar stock of A-286, INCO 901, Astroloy, Waspaloy, B1900, IN-100, and titanium alloys 8-1-1 and 6-4 were produced by controlled processing so that they behaved in a superplastic manner over a fairly wide temperature range if the strain rates were properly controlled (Figures 5, 6, and 7).

The next logical step was to forge these alloys isothermally in the temperature range and at the strain rate where they demonstrated superplastic behavior. It was found that billet stock previously made superplastic could be forged isothermally into complex configurations using very low unit pressures (Figure 8). It was also found that billet stock which did not demonstrate superplastic behavior could be made superplastic by forging isothermally at the proper temperature and strain rate, so that complex configurations could be produced at the finish of the forge cycle.

Both IN-100 and Astroloy have been forged starting with a powder product, as well as from cast ingots. Superplastic forging stock has been produced by various methods such as extrusion, hot rolling, and press forging.

This controlled isothermal forging practice, which utilizes the superplastic behavior of the alloy being forged, has been given the name *Gatorizing*[TM] by Pratt & Whitney Aircraft and is covered by U.S. Patent 3,519,503.

Some of the advantages of this process are as follows:

1. The ultrahigh-strength, low-ductility nickel-base super-alloys developed for cast turbine blades can be forged with extreme ease. In fact, the more difficult such an alloy is to forge conventionally, the easier it is to forge by this new process.

2. Superalloys can be forged to extremely close tolerances without risk of cracking. This allows a large reduction in forging input weight with attendant reduction of machining costs. As-forged surfaces of many forgings are sufficiently smooth to exclude final machining.

3. Forgings of complex configuration that would be difficult—if not impossible—to forge with uniform mechanical properties by conventional means can be forged with relative ease to provide uniform properties with the *Gatorizing* forging process.

4. Forging pressures required with this process are substantially less than those necessary in conventional forging, thus allowing the use of lighter forging equipment.

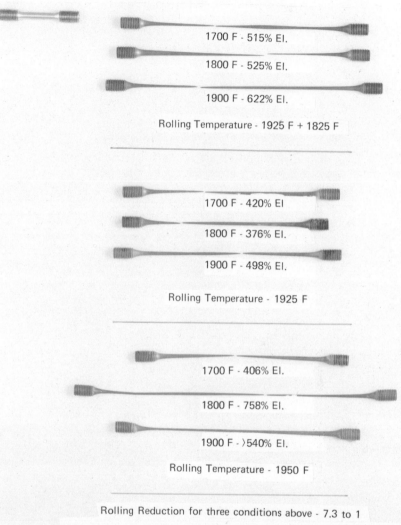

1700 F - 515% El.

1800 F - 525% El.

1900 F - 622% El.

Rolling Temperature - 1925 F + 1825 F

1700 F - 420% El

1800 F - 376% El.

1900 F - 498% El.

Rolling Temperature - 1925 F

1700 F - 406% El.

1800 F - 758% El.

1900 F - >540% El.

Rolling Temperature - 1950 F

Rolling Reduction for three conditions above - 7.3 to 1

Note: Temperatures shown under each specimen are test temperatures.

RUPTURE

Figure 5. Superplastic Astroloy billet.

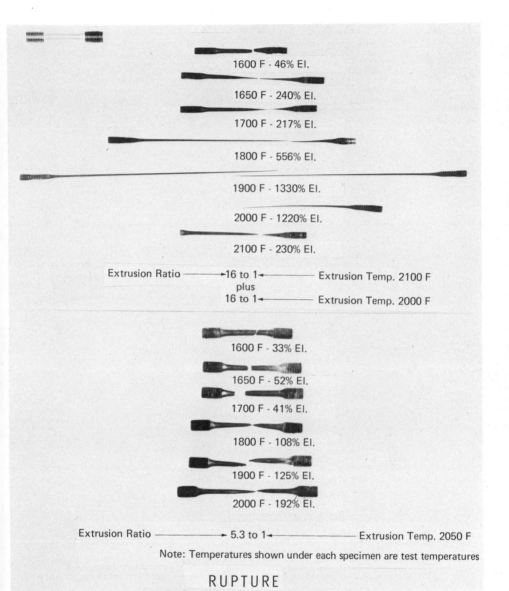

1600 F - 46% El.

1650 F - 240% El.

1700 F - 217% El.

1800 F - 556% El.

1900 F - 1330% El.

2000 F - 1220% El.

2100 F - 230% El.

Extrusion Ratio ——————►16 to 1◄—————— Extrusion Temp. 2100 F
plus
16 to 1◄—————— Extrusion Temp. 2000 F

1600 F - 33% El.

1650 F - 52% El.

1700 F - 41% El.

1800 F - 108% El.

1900 F - 125% El.

2000 F - 192% El.

Extrusion Ratio ——————► 5.3 to 1◄—————— Extrusion Temp. 2050 F

Note: Temperatures shown under each specimen are test temperatures

RUPTURE

Figure 6. Superplastic IN-100 billet.

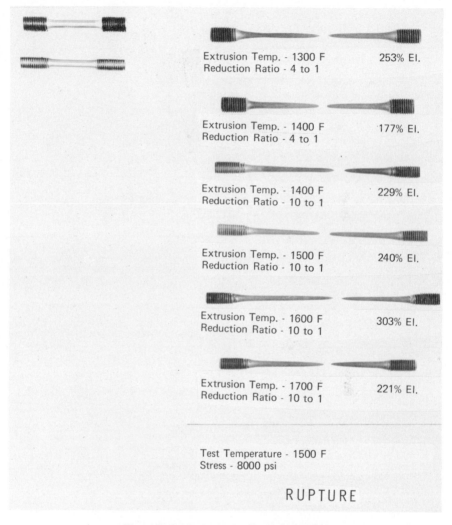

Figure 7. Superplastic titanium 8–1–1 billet.

The scale-up of the "all inert" powder process has involved the production of IN-100 compacted powder billets which were extruded to bar suitable for forging into disks.

The IN-100 powder was produced, collected, and compacted into 5-inch-diameter by 8-inch-high billets weighing 45 pounds, using the "all inert" P/M process (Figure 9). A theoretical density of 99 percent was achieved. These billets were "canned" and extruded to 2.25-inch-diameter bar stock. Metallo-

Figure 8. *Gatorized*[TM] disks.

Mag 100X

Figure 9. Hot pressed IN-100 Powder billet. Right, 100X, reduced to 55% of original.

graphic examination of the stock revealed an extremely uniform structure with no voids or defects (Figure 10).

The bar stock was forged by *Gatorizing* into 5-inch-diameter disks, which were used for heat-treatment and mechanical property evaluation (Figure 11).

A brief review is needed to show the persistence of macrosegregation from a cast ingot through billet conversion into the finished disk forging, as opposed to its absence in a powder product. Figure 12 is the product of a controlled structure 12-inch-diameter Astroloy ingot reduced to a 3.5-inch-diameter superplastic billet and forged by *Gatorizing* into the disk shown with a reduction of approximately 80 percent of the billet height. The flow lines are bands of segregation. The severity of the segregation increases with increasing ingot size, as shown by Figure 13, which is the product of a controlled structure 20-inch-diameter Astroloy ingot reduced to an 8-inch-diameter superplastic billet and forged by

Mag 100x

Figure 10. IN-100 2.25-inch-diameter bar stock from powder billet has extremely uniform microstructure. Right, 100X, reduced to 60% of original.

Figure 11. IN-100 flat disk forging from "all inert" powder.

Gatorizing into the disk shown, with a reduction of approximately 80 percent of the billet height.

Figure 14 shows a small IN-100 disk forging from "all inert" powder bar of 2.25 inches extruded from a 5-inch-diameter compacted billet. This disk is also etched to show flow lines. The structural uniformity and absence of macro-segregation is unprecedented in Pratt & Whitney Aircraft's experience with nickel-base superalloy forgings. The IN-100 alloy, being stronger than Astroloy, is even more segregation-prone than Astroloy, which further emphasizes the benefits of eliminating macrosegregation through the use of "all inert" powder.

The parameters established for the 5-inch-diameter compacted billets were used to produce 10.5-inch-diameter billets (Figure 15) which were extruded to

Mag 1X

Figure 12. Diametrical macroslice of an as-forged Astroloy turbine disk—product of 12-inch-diameter ingot, reduced to 50% of original.

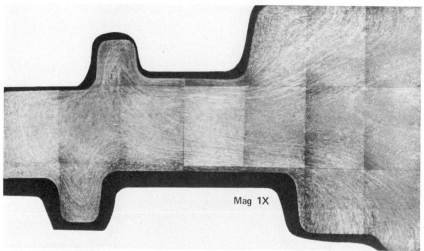

Figure 13. Radial macroslice of an as forged Astroloy turbine disk–product of 20-inch-diameter ingot, reduced to 50% of original.

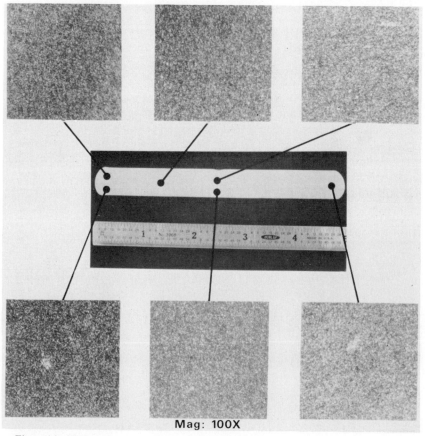

Figure 14. Diametrical macroslice of as-forged IN-100 pancake–product of 5-inch-diameter powder billet. 100X, reduced to 70% of original.

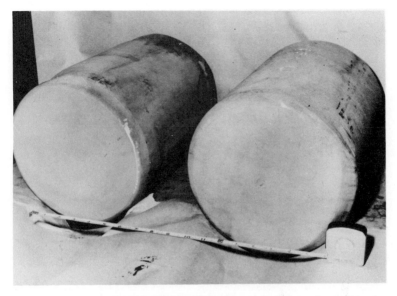

Figure 15. 10.5-inch-diameter compacted IN-100 powder billets prior to extrusion.

4.7-inch-diameter superplastic forging stock (Figure 16). The same uniformity from edge to center and from one end of the extrusion to the other was found in these larger extrusions as was seen in the 2.25-inch extrusions.

Several flat disks were forged by *Gatorizing* from the 4.7-inch billet, the largest weighing approximately 70 pounds. Metallographic examination of the surface of one of these disks (Figure 17) shows a degree of structural uniformity and absence of segregation that is unique for nickel-base superalloys. The cross section of the same disk (Figure 18) shows similar uniformity and freedom from die lock and segregation as shown by the surface examination.

The mechanical properties obtained from these IN-100 forgings are superior to the properties obtained from Astroloy disk forgings presently used in advanced production engines (Figures 19, 20, and 21). The tensile properties, creep life, and stress-rupture life are compared to Astroloy minimum curves. The LCF life is compared to Waspaloy and Astroloy LCF life in Figure 22.

The scale-up of IN-100 "all inert" powder has progressed to forging stock which is large enough to produce *Gatorized* engine disk forgings of approximately 150 pounds (Figure 23). This material shows the same uniformity of structure and freedom from alloy segregation as the smaller diameter forging stock. The IN-100 billet stock has been produced by hot compaction followed by extrusion, as well as by a one-step compaction-extrusion process. Material from each process has produced satisfactory compressor and turbine disks. Well over 100

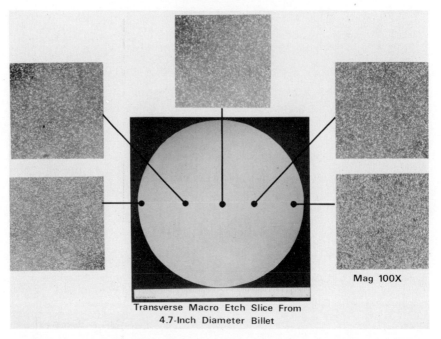

Mag 100X

Transverse Macro Etch Slice From
4.7-Inch Diameter Billet

Figure 16. Extruded IN-100 billet has exceptional uniformity of structure and absence of segregation. 100X, reduced to 55% of original.

compressor and turbine disks for an advanced jet engine have been forged by the *Gatorizing* process from powder-metal IN-100 billets.

The input weight to make a particular Astroloy engine disk by the *Gatorizing* process was 160 pounds. To make the same disk by conventional forging, one forger used an input weight of 390 pounds; another, one of 440 pounds.

Summary

Work performed by Pratt & Whitney Aircraft has conclusively demonstrated the applicability of "all inert" powder metallurgy to the manufacture of disks from the most advanced nickel-base superalloys for high-performance jet engines. Without the use of this powder, such alloys would be unusable or marginal at best because of segregation inherent to their chemical compositions. "All inert" powder when combined with the *Gatorizing* forging process yields a disk product which is unique for its extreme structural uniformity and uniformity of mechan-

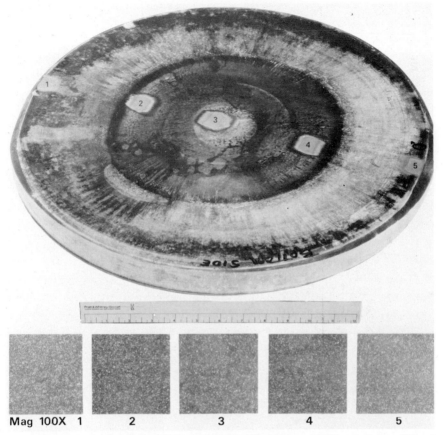

Mag 100X   1   2   3   4   5

Figure 17. Flat disk forging from 4.7-inch-diameter billet. 100X, reduced to 65% of original.

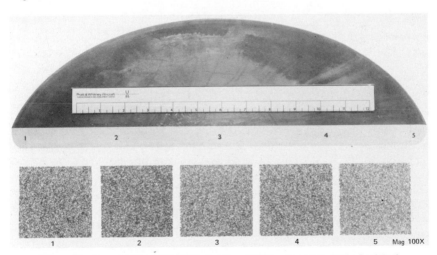

Figure 18. Cross section of flat disk forging. 100X, reduced to 50% of original.

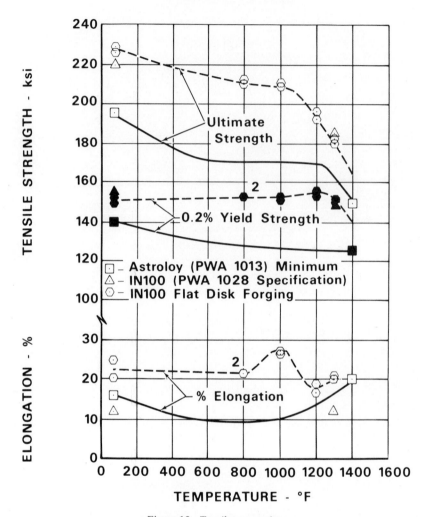

Figure 19. Tensile properties.

ical properties. Although the present cost of "all inert" powder billet exceeds that of conventionally produced billet, it is believed that powder billet will be competitive in the future when there is a reasonable volume of superalloy usage. When this happens, the savings inherent in the *Gatorizing* forging process will result in an engine product which will be less costly than a similar part made from the same superalloy using conventional melting and forging techniques.

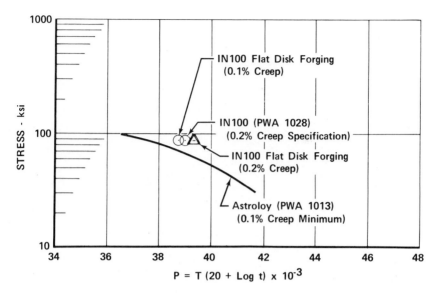

Figure 20. Creep life (Larson-Miller parameter plot).

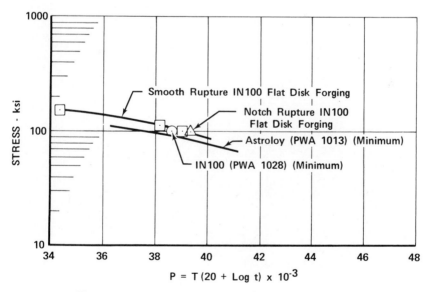

Figure 21. Stress rupture life (Larson-Miller parameter plot).

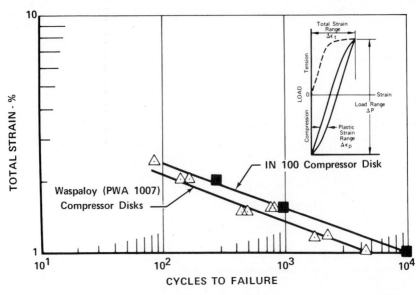

Figure 22. Low-cycle fatigue *Gatorized* IN-100 vs. Waspaloy constant strain 1200°F.

Figure 23. *Gatorized* engine disk forging.

The utilization of "all inert" powder metallurgy also opens up potential for the development and use of higher temperature strength superalloys than those in use today. Because powder alleviates segregation, it is conceivable that higher hardener contents can be tolerated in existing alloy systems for making acceptable high-performance engine hardware. In addition, powder affords the opportunity for dispersion hardening to effect improvements in existing alloy systems or the development of new ones.

15. Complex Superalloy Shapes

A. S. BUFFERD

Whittaker Nuclear Metals
West Concord, Massachusetts

ABSTRACT

Complex superalloy shapes prepared by powder-metallurgy techniques rely on the fine grain size and chemical homogeneity advantages of powder as compared to cast and wrought material. Superalloy powders provide the basis for advanced metalworking techniques such as filled billet extrusion and rolling as well as for superplastic deformation in forging and drawing. Attention is directed not only to the metalworking process but to the influence of chemistry and heat treatment in developing engineering properties in complex shapes prepared with superalloy powders.

Introduction

The last decade has provided for increased activities in the development of alloy metal powders and the further consideration of wrought shapes from powders as compared to cast ingots. These activities were proceeding while, at the same time, nickel superalloys became more complex in both their chemistry and processing in order to meet the requirements of a most demanding system, the turbine engine power plant.

Much of the early work in this area considered the use of elemental powders to form the desired alloys or utilized alloy powders prepared by water-atomization techniques. Difficulties in evaluating the results of processing and property studies because of incomplete alloying or high interstitial contents such as oxygen have been overcome with the current availability of high-quality superalloy powders prepared by inert gas atomization or the rotating-electrode process.

Another feature of recent efforts with superalloy powders is the almost complete absence of process studies directed to cold pressing and sintering. This is directly related to the intrinsic high strength of the alloys and to the characteristics of inert gas-atomized or REP powders which are spherical in shape and therefore virtually incompressible at room temperature. As a result, the current

303

activities in superalloy powder metallurgy are strongly centered on hot consolidation processes which produce a semifinished shape or a preform for subsequent metalworking.

A process of significant importance in this regard is hot isostatic pressing, which has been described elsewhere in this volume [1]. The hot extrusion process currently provides for the bulk of superalloy powder processing, either in terms of mill forms such as bar and wire or for the generation of complex shapes.

Hot Extrusion Process

Earlier work in describing the hot extrusion of metal powders was mainly concerned with pure metals such as chromium or beryllium and dispersion-strengthened materials [2,3]. However, in both the previous work with pure metals and current efforts with superalloys, the hot extrusion process has shown the capability for combining hot compacting and hot mechanical working to yield a fully dense wrought material. In particular, it is the frictional forces in the hot extrusion process which yield a shear component that gives rise to inter-particle shearing, a breakdown in prior particle boundaries, and promote inter-particle bonding.

The most important technique for the hot extrusion of superalloy powders is the packing of the powders into a metallic shell or "can," which is extruded with the superalloy. When the can is filled with powder to its tap density (62 percent of theoretical for spherical powders) without prior compaction, the extrusion is referred to as a direct powder extrusion. The can is provided with an evacuation tube on an end-plate such that the powder may be outgassed at room and elevated temperatures and sealed before heating for extrusion. This effectively eliminates atmosphere contamination of the alloys during hot extrusion.

The final shape of the densified powder is determined in part by the way in which the can deforms relative to the powder during the initial upsetting which precedes extrusion through the die. In general, for a given can configuration (can-wall thickness to can length) there is a greater tendency for can-wall buckling with a lower density of packing of the powder. This condition is illustrated in Figure 1, and results not only in poor shape control but in a significantly lower yield of extruded metal. The spherical shape of superalloy powders, which does not provide for powder bridging and results in a high packing density, tends to minimize this difficulty in contrast to other types of metal powders.

When can buckling does occur, this is correctable either by increasing the ratio of can-wall thickness to can length or by incorporation of the penetrator tech-

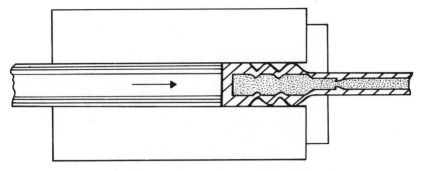

Figure 1. Folding of metal can with loosely packed powders.

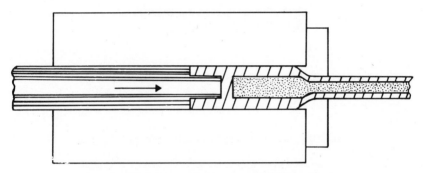

Figure 2. Penetrator technique to avoid folding of can.

nique shown in Figure 2. The penetrator cuts through the top of the can and densifies the powder within the can prior to upsetting of the can.

Other dimensional variations of the extruded powder are characteristic of the process of coextrusion and result from the nature of material flow during extrusion. Some characteristic variations are shown in Figure 3, and must be taken into account in considering the yield of material resulting from the hot extrusion of metal powders.

In addition to the considerations of can geometry, the canning material must satisfy a number of requirements. First, and probably most important, the stiffness of the can material at the extrusion temperature should be as close as possible to that of the powder to be extruded. If there are significant differences in the deformation characteristics of the can material and the superalloy powder, defects such as those shown in Figure 4 may arise, and further result in cracks extending into the consolidated powder core. The can material should not react with the superalloy powders at the extrusion temperature, and should be removable by chemical pickling or mechanical means from the consolidated powder.

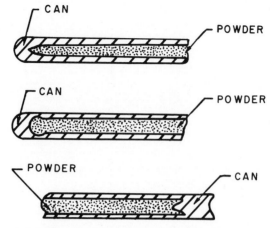

Figure 3. Characteristic dimensional variations. Top: Can thickening at the front. Center: "Dogboning" of the consolidated powder at the front. Bottom: "Fishtailing" of the consolidated powder at the rear.

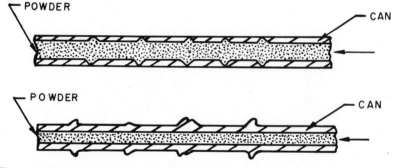

Figure 4. Defects resulting from can-core differences in deformability. Top: Can stiffer than core. Bottom: Core stiffer than can.

Figure 5 shows some typical extrusion constants for superalloys and two candidate canning materials over a range of temperatures. The extrusion constant, K, is derived from experimental measurements by the relation:

$$K = \frac{F}{A \, \ln R}$$

where

F = extrusion force in tons

A = cross-sectional area of the billet in square inches

R = extrusion reduction ratio, which in turn is the ratio of the billet cross-sectional area to the extruded bar cross-sectional area.

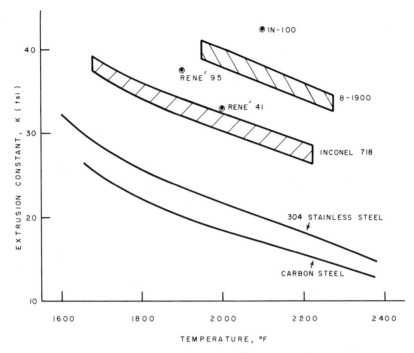

Figure 5. Extrusion constants of superalloys and canning materials.

In general, stainless steel is utilized as the canning material for the direct powder extrusion of superalloys and must be removed by mechanical means from the extruded bar. While material yield losses due to differences in deformation may be adjusted for by selection of the extrusion temperature, there are practical limits to tooling stresses which further restrict process flexibility.

However, the most significant factor governing the extrusion of superalloy powders is an understanding of the interplay between the extrusion variables of temperature and reduction ratio on the resultant microstructure and properties of each superalloy composition of interest. The same procedures do not apply for all alloys.

Two superalloys, Inconel 718 and René 41, were extruded at 2000°F with a 12:1 reduction ratio. In both cases, a fully dense wrought bar with good material yield resulted. However, Figure 6 shows the presence of prior powder-particle boundaries in the René 41 alloy and not in the Inconel 718. This is further illustrated in Figure 7, which compares the resultant grain growth in each alloy after solution treatment at 2200°F and aging. It is to be noted that while grain growth proceeded unhindered in the Inconel 718, the prior particle boundaries in the René 41 blocked grain growth. This effect was evident with

Figure 6. Superalloys Extruded at 2000°F and 12:1 Reduction.   (a) Inconel 718; (b) René 41. Longitudinal. 250X.

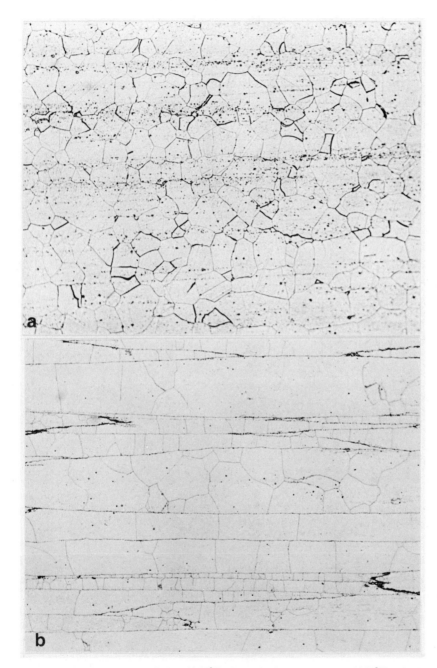

Figure 7. Superalloys extruded at 2000°F after solution treatment at 2200°F and aging. (a) Inconel 718; (b) René 41. Longitudinal. 250X.

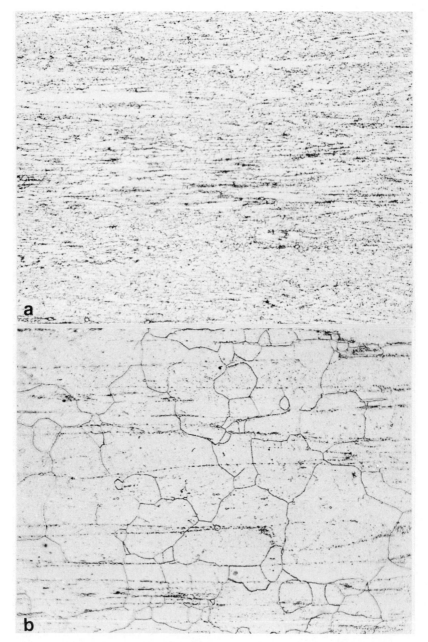

Figure 8. Longitudinal sections of René 41 processed by extrusion at 1850°F and 12:1 reduction. (a) As-extruded; (b) solution treated at 2200°F and aged. 250X.

other reduction ratios above 12:1 as long as the extrusion temperature was 2000°F.

It has been hypothesized that this effect noted with René 41 and other alloys is due to a temperature-sensitive transformation which results in the formation of carbide phases at the particle boundaries, and that this may be overcome by an appropriate modification of the extrusion temperature for each alloy [4]. Figure 8 shows structures confirming this hypothesis for René 41 as a result of an extrusion at 1850°F with a 12:1 reduction ratio. A comparison of the solution-treated and aged grain structures of René 41 in Figures 7 and 8 emphasizes this influence of extrusion temperature.

While metallography indicated the importance of the extrusion temperature in eliminating prior particle boundaries, selected mechanical property tests did not indicate as strong an influence of the extrusion temperature on elevated temperature properties. Table I compares the tensile properties at 1400°F and the stress-rupture properties at 1650°F for René 41, with the only process difference being the extrusion temperature. The only significant property difference is the reduction in area in the 1400°F tensile test, being slightly higher with the lower extrusion temperature. While these tests were conducted on longitudinal sections, other data on Udimet 700 shows this correlation with metallographic structure and transverse mechanical properties.

TABLE I

Properties of René 41 Extruded at 12:1 Reduction,
Then Solution-Treated and Aged

| | Tensile at 1400°F | | | | Rupture Hours at 1650°F, 25 ksi |
	U.T.S. (ksi)	Y.S. (ksi)	Elong. (%)	R.A. (%)	
1850°F Extrusion	144	136	22	35	12
2000°F Extrusion	143	134	21	27	12
AMS 5713B Spec.	135	105	5	8	–

Other processing interests are more directed to the grain structure of extruded superalloy powders. It is known that superalloys containing an extremely fine-grained structure are candidates for superplastic deformation [5]. Superalloy powder, consolidated under conditions to promote this fine grain size, is the basis of a new technology that will allow for the superplastic forging of turbine engine parts [6]. In addition, attention is being directed to utilizing the super-plasticity concept with superalloys in other areas such as hard facing and brazing alloy wire manufacture with the cobalt- and nickel-base alloys and the precision drawing of superalloy structural shapes.

Therefore, in considering the extrusion of superalloy powder to bar form, the selection of the specific extrusion conditions will vary, not only from alloy to alloy but also in regard to the further processing considered for the material. In the case of extruded powder bar to be forged superplastically into a turbine engine disk, the very fine grain size required would suggest a different set of conditions than if the extruded powder bar were to be utilized directly after heat treatment as high-temperature fastener material.

With the awareness of the direct influence of the extrusion conditions on the properties of superalloys prepared from powders, it has been possible to consider the use of such powders together with the "filled billet" extrusion concept to generate complex shapes [7].

Filled Billet Extrusion

The filled billet extrusion technique can be used to produce shapes of either complex cross section or of a hollow nature. This is especially important where tooling technology does not allow for direct fabrication of the cross section. In addition to hollow vanes, cross sections such as those shown in Figure 9 are currently of interest in the superalloy field.

The filled billet extrusion technique initially was based on solid materials without consideration of superalloys or of direct powder extrusion. With this technique (Figure 10), an enlarged replica of the desired shape is produced within a round extrusion billet and surrounded by a filler material. After extrusion under streamlined flow conditions to a round bar, the filler material is removed to yield the desired shape.

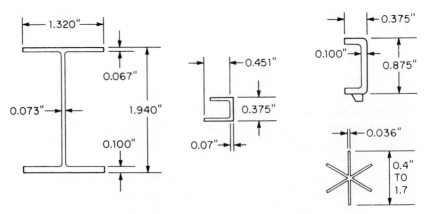

Figure 9. Typical superalloy sections.

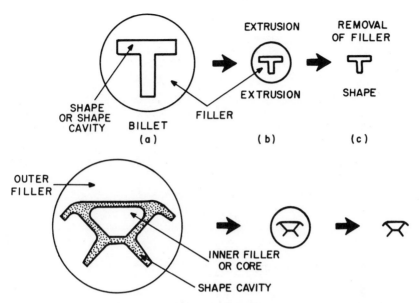

Figure 10. Outline of filled billet process.

In considering the preparation of a filled billet, the machining of the enlarged replica of the desired shape and a matching filler from solids is replaced by only machining the matching filler and filling the shape cavity with superalloy powder which is consolidated during an extrusion sequence designed also to provide the structure and properties required in the shape after heat treatment.

When both components of the filled billet are 100-percent dense (as from machined solids), the enlarged replica is reduced in linear cross-section dimensions by the square root of the extrusion reduction ratio and elongated by a factor equal to the reduction ratio. However, when only the filler is 100-percent dense and the enlarged replica is 62-percent-dense spherical powder, a dimensional correction must be made for the enlarged replica dimensions, since they will change relative to the over-all billet during the early stages of extrusion (*i.e.*, during the initial upsetting operation). The dimensional correction factor is based on the apparent density of the powder and the relative cross-sectional areas of the shape and the filler, which will, of course, vary from one shape to another.

This has resulted in a good degree of predictability and control of shape dimensions such that Figure 11 defines the current limits and tolerances for complex superalloy shapes prepared by the filled billet extrusion technique.

Through a coupling of this technique with a knowledge of the influence of the extrusion variables on properties of direct powder extruded superalloys, a design outline for dimensions and properties can be established:

1. Define objective shape dimensions and properties;

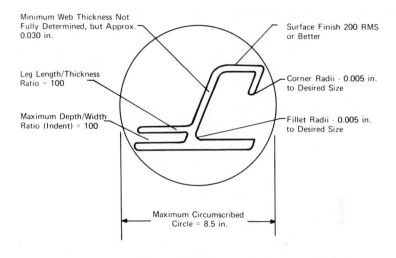

Minimum Web Thickness Not Fully Determined, but Approx. 0.030 in.

Surface Finish 200 RMS or Better

Leg Length/Thickness Ratio = 100

Corner Radii · 0.005 in. to Desired Size

Maximum Depth/Width Ratio (Indent) = 100

Fillet Radii · 0.005 in. to Desired Size

Maximum Circumscribed Circle = 8.5 in.

TOLERANCES			
Height and Width of Legs	2 in. and up ± 10%	Angles	± 2°
Web Thickness	± 0.010 in.	Transverse Flatness Max.	0.020 in./in.
Corner Radii	± 0.030 in.	Fillet Radii	± 0.030 in.

Figure 11. Limits and tolerances for superalloy shapes by filled billet extrusion.

2. Determine as-extruded bar size from a circumscribed circle of the objective shape cross section;

3. Select extrusion temperature and reduction ratio to yield objective properties;

4. Apply the dimensional correction factor in the billet design;

5. Machine filler.

At this stage, all further processing is comparable to that of a direct powder extrusion for bar except that this approach has a can shaped on the inside to define a complex shape, and the canning material selected is usually carbon steel for easy removal after extrusion.

Figure 12 shows some of the shapes, including hollows, which have been prepared from superalloy powders. Materials include Inconel 718, René 41, and Inconel 625. Table II presents mechanical property measurements of samples taken from an Inconel 718 complex shape prepared by direct powder extrusion and the filled billet technique. The results demonstrated the concept of combining the filled billet technique with control of the direct powder extrusion

Figure 12. Superalloy shapes prepared by the filled billet technique with superalloy powder.

TABLE II

Tensile Properties at 1200° F of Inconel 718 Direct
Powder Extrusion-Complex Shape as Heat-Treated

Specimen	U.T.S. (ksi)	Y.S. (ksi)	Elong. (%)
Longitudinal			
AMS 5663B	145	125	12
Flange section	157	136	17
Web section	158	130	28
Transverse			
AMS 5663B	140	125	6
Web section	155	135	10

parameters to yield properties required with a complex shape. This concept is currently being applied to other superalloys and geometrical configurations.

In addition to the extrusion technique to process complex shapes from powder, other work has been indicated with a similar approach in rolling [8]. Further efforts have been suggested to produce the complex shape with a very fine grain size such that superplastic drawing may be utilized for extremely thin sections or improved tolerances.

Summary

The hot extrusion of supperalloy powders has been described in terms of not only producing a fully dense wrought structure but also the need to consider the influence of extrusion temperature and reduction ratio on the resulting structure and properties. This approach has then been extended to the fabrication of complex shapes where structure control can result in the required properties after heat treatment or in further shape modification through superplastic deformation.

References

1. Hanes, H.D., "Hot Isostatic Compaction of High Performance Materials," *Powder Metallurgy for High-Performance Applications*, J. J. Burke and V. Weiss, eds., Syracuse, N.Y.: Syracuse University Press (1972).
2. Lowenstein, P., Aronin, L.R. and Geary, A.L., "Hot Extrusion of Metal Powders," *Powder Metallurgy*, Werner Leszynski, ed., New York: Interscience Publishers (1961), 563.
3. Gardner, N.R., "The Extrusion of Metal Powders," *Progress in Powder Metallurgy*, Vol. 19, New York: MPIF (1963), 135.
4. Friedman, G.I., Personal communication.
5. Headley, T.J., Kalish, D. and Underwood, E.E., "The Current Status of Applied Superplasticity," *Ultrafine-Grain Ceramics*, J.J. Burke, N.L. Reed, and V. Weiss, eds., Syracuse, N.Y.: Syracuse University Press (1970), 325.
6. U.S. Patent No. 3,519,503, July 7, 1970, *Fabrication Method for the High Temperature Alloys*. J.B. Moore, J. Tequesta, and R.L. Athey to United Aircraft Corporation, East Hartford, Connecticut.
7. Gorecki, T.A. and Friedman, G.I., "Extruded Structural Shapes for Superalloy Powders," WESTEC Conference, Los Angeles, California, March 1971.
8. U.S. Patent No. 3,531,848, October 6, 1970, *Fabrication of Integral Structures*, P.J. Gripshover and H.D. Hanes to The Battelle Development Corporation, Columbus, Ohio.

16. High-Speed Tool Steels by Particle Metallurgy

E. J. DULIS
Crucible Materials Research Center
Pittsburgh, Pennsylvania

ABSTRACT

One relatively recent use of powder metallurgy for high-performance applications is in the high-speed tool steel field. A long-standing problem in conventionally cast tool steel ingots involves segregation related to the slow cooling of ingots. This problem is overcome by employing prealloyed powder and compacting to large sections that are then processed on conventional mill facilities to standard high-speed steel product forms. The improved microstructure is found to be directly related to significant improvements in important properties of P/M tool steels. Such properties as improved cutting-tool life, out-of-roundness distortion after hardening heat treatment, and grindability, are significantly better in P/M tool steels. The improved conventional grades of high-speed tool steels made by P/M are believed to be the forerunners to the introduction of vastly improved tool materials made by the P/M process.

Introduction

Upgrading of high-speed tool steels through eliminating or minimizing carbide segregation has been an objective of tool-steel producers around the world for the past seventy years. Carbide segregation is normal to the slow cooling involved in solidification of large ingots. This segregation, particularly in large-sized tools, adversely affects important tool properties.

To overcome the carbide segregation problem, the powder-metallurgy process has been found to be an excellent approach and experimental small-sized sections were produced by various organizations. However, the obstacles to producing P/M high-speed tool steels commercially were (1) unavailability of good-quality powder, and (2) difficulty in making fully dense large section sizes. Employing the relatively new technology in vacuum and high-pressure systems developed for other fields, we developed a new process for making P/M high-speed tool steels [1,2,3].

Our process, illustrated in Figure 1, is now at the stage where we are produc-

317

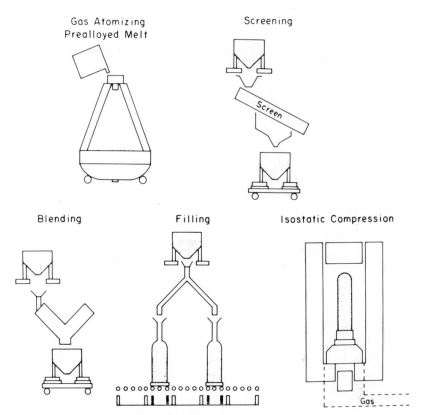

Figure 1. Particle-metallurgy processing at Crucible Specialty Metals Division, Colt Industries.

ing commercially all the popular types of high-speed tool steels in compacts weighing up to 1,400 lbs each. These compacts are then further processed to various product forms on conventional steel-mill facilities.

Compared to conventionally cast-and-wrought high-speed tool steel, P/M products exhibit a uniform distribution of fine carbides. The microstructures of conventionally cast and as-atomized particles of high-speed tool steel M2S show the initial inherent structural differences (Figure 2). In contrast to the segregation found in hot-worked bars of conventional steel, P/M product maintains a fine carbide size and homogeneous distribution regardless of the size of the mill product, as shown on Figure 3. The significance of fine uniformly distributed primary carbide phase is in its relationship to grain size, tool life, grindability, and out-of-roundness distortion after hardening heat treatment.

The grain-size and carbide-size relationship in both P/M and conventional

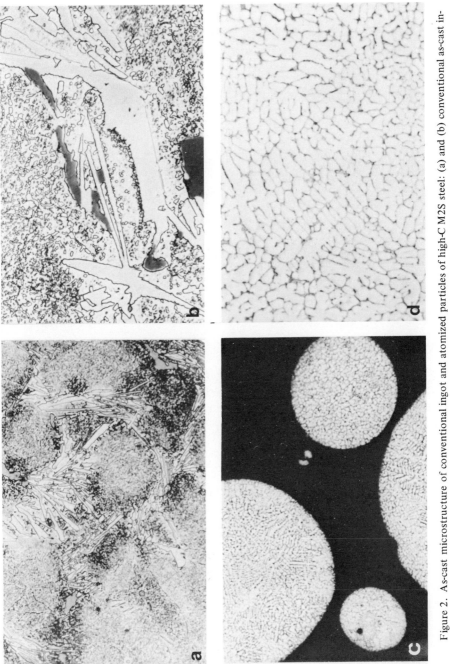

Figure 2. As-cast microstructure of conventional ingot and atomized particles of high-C M2S steel: (a) and (b) conventional as-cast ingot; (c) and (d) as-atomized particles. Magnification of (a) and (c), 250X; (b) and (d), 1000X.

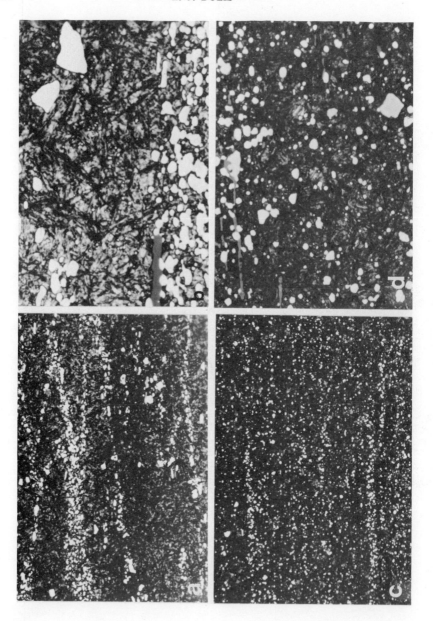

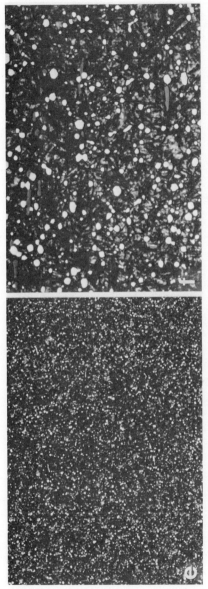

Figure 3. Carbide size and distribution in conventional high-C M2S bars and in P/M products. (a) and (b) conventional 5" (127 mm) bar (mid radius); carbide size, mean 4 μm, max. 24 μm. (c) and (d) conventional 3/4" (19 mm) bar; carbide size, mean 3 μm, max. 16 μm. (e) and (f) P/M 5" (127 mm) bar; carbide size, mean 2 μm, max. 4 μm. Magnification of (a), (c), and (e), 250X; (b), (d), and (f), 1000X.

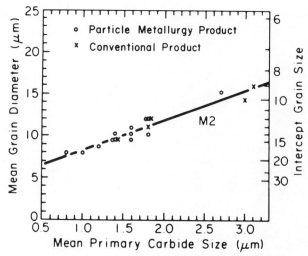

Figure 4. Effect of primary carbide size on the austenite grain size in high-speed steel (Neumeyer and Kasak [4]).

products show that the finer the primary carbides, the smaller the resulting grain size after hardening heat treatment, as seen in Figure 4 [4]. The inherently smaller carbide and grain size of P/M, as compared to conventionally produced steel, is clearly evident.

The relationship of grain size with bend fracture strength and intermittent cutting-tool life shows that these properties are improved as the grain size becomes smaller (Figure 5). Bend fracture strength is a commonly accepted laboratory measure of high-speed tool steel toughness and performance. Intermittent cutting-tool life provides an actual measure of the ability of the tool to withstand the continuous series of impacts involved in many cutting operations.

An unexpected but very significant advantage of P/M-produced material is the improvement in grindability. A measure of grindability is the volume of hardened steel removed as a function of the volume of refractory grinding wheel lost (Figure 6). Comparing the slopes of grinding-wheel loss versus metal volume removed for conventional high-carbon M2S and P/M high-carbon M2S, both heat-treated to R_c 65, shows that the P/M steel has a 3:1 advantage. The findings of improved grindability and of improved tool life may seem contradictory, inasmuch as one would expect improved wear resistance in cutting to correspond to increased wear resistance in grinding, i.e., poorer grindability. Cutting performance involves mild and primarily adhesive wear. Under these conditions, a typical P/M product with fine, uniformly dispersed carbides is more wear resistant than conventional product. However, grinding involves severe and primarily abrasive wear; under these conditions, the fine carbides offer little resistance to removal by the grinding wheel in comparison to the coarse carbides

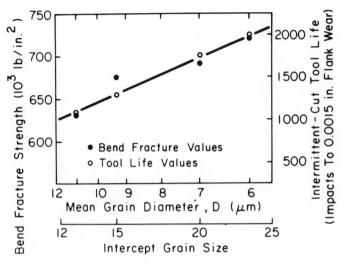

Figure 5. Effect of austenite grain size on bend fracture strength and intermittent-cut tool performance of M2. Cutting Speed: 60 ft/min (Neumeyer and Kasak [4]).

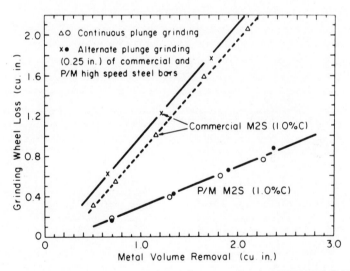

Figure 6. Grinding ratios (reciprocal slopes) for conventional and P/M M2S (1.0% C).

of the conventional product. A comparison of the carbide size differences between the conventional and P/M high-speed steels shows that the MC-type vanadium carbide size differences are even more pronounced than the total primary carbides (Figures 7, 8, and 9). The measured primary vanadium MC carbide distribution indicates (1) a mean size of 1 μm and maximum of 2 μm for

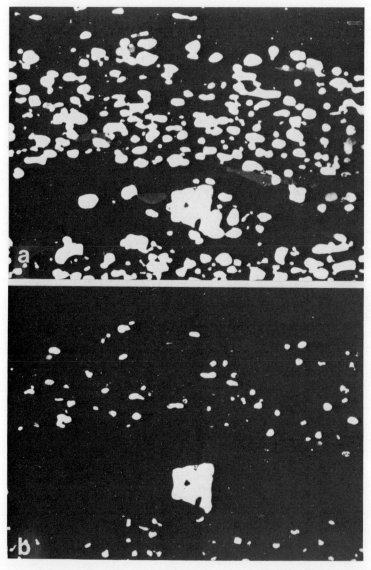

Figure 7. Primary carbides in a 5-inch-diameter bar of P/M M2S (1.0% C). 1000X.
(a) Etched to show size and distribution of M_6C and MC carbides (picral etchant).
(b) Same area as (a), etched to reveal only MC carbides (picral and Murakami's etchants).

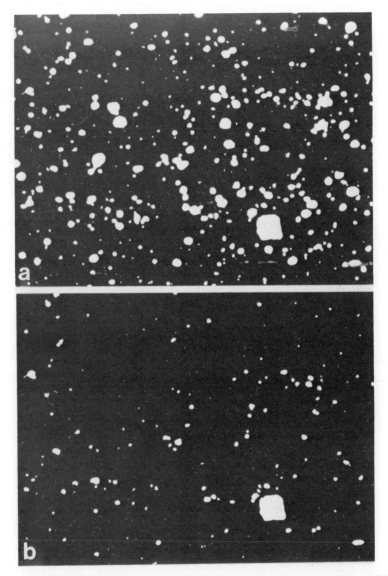

Figure 8. Primary carbides in a 3/4-inch-square bar of conventional M2S (1.0% C). 1000X. (a) Etched to show size and distribution of M_6C and MC carbides (picral etchant). (b) Same area as (a), etched to reveal only MC carbides (picral and Murakami's etchants).

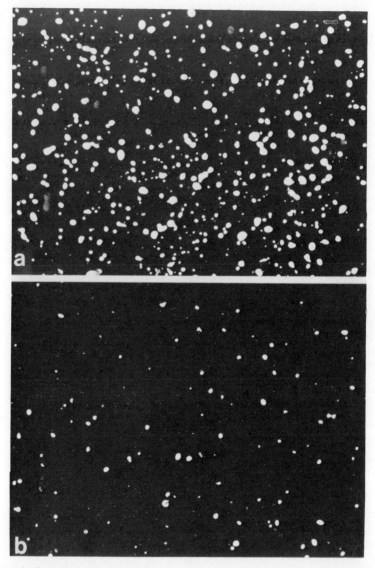

Figure 9. Typical primary carbides in a 5-inch-diameter bar of conventional M2S (1.0% C).   1000X.   (a) Etched to show size and distribution of M_6C and MC carbides (picral etchant).   (b) Same area as (a), etched to reveal only MC carbides (picral and Murakami's etchants).

Material	Range of Reported Knoop Hardness Values	Typical Average Knoop Hardness
High Speed Steel (Rc 64-70)	820 — 1100	900
W-Mo-Fe Carbides (M_6C Type)	1400 — 1650	1500
Vanadium Carbides (MC Type)	2300 — 2800	2600
Al_2O_3 (Corundum) Grinding Wheel	1900 — 2900	2400
Silicon Carbide Grinding Wheel	——	2500

567-71

Figure 10. Hardness of major components in the grinding of a high-speed steel.

P/M steel, and (2) a mean size of 3 μm and maximum of 16 μm for conventional 3/4-inch-diameter bar. For the 5-inch-diameter conventional bar, the mean is 4 μm and maximum is 24 μm. The differences in hardness of the steel matrix, carbides, and grinding-wheel materials shows MC vanadium carbide to have a higher hardness and the matrix and M_6C carbides a lower hardness than the grinding-wheel abrasive materials (Figure 10). Thus, the large MC carbides will "dress" the grinding wheels until they are removed from the surface. The fine MC carbides are more readily removed from the surface during grinding.

A very significant advantage of P/M high-speed steels is the substantially reduced out-of-roundness or improved uniformity of size change that occurs during the final hardening heat treatment of machined-to-shape large tools. A comparison of the out-of-roundness of 7.6-inch-diameter bars of conventional (0.0020 inch) and P/M (0.0003 inch) high-carbon M2S (Figure 11), shows an approximate 7:1 advantage for the P/M material. Also, the P/M steel did not have the rectangular shape exhibited for the conventional steel. In general, three- to tenfold improvements have been found in large P/M rounds as compared to conventional rounds. This improvement is to a large degree related to the uniform microstructure and random crystallographic orientation of the P/M products, as compared with the highly segregated microstructure of the conventionally produced products. The significance of increased uniformity of size change after final heat treatment is that the need for grinding can be eliminated; this leads to a significant reduction of the manufacturing cost. Even if grinding is required, the probability of surface damage would be minimized. Depending on the degree and care in grinding of the hardened tool, the surface damage can result in decreased performance or scrapping of the final tool.

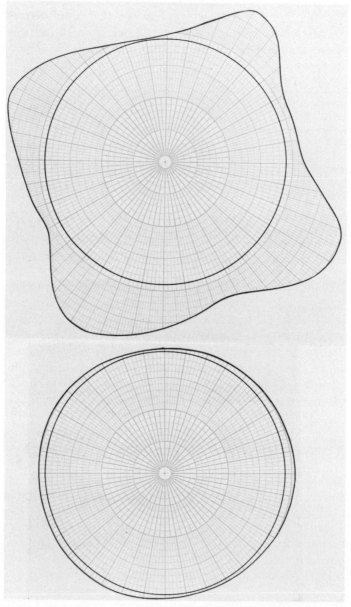

Figure 11. Out-of-roundness distortion after hardening heat treatment of P/M and conventional high-carbon M2S. Top: Conventional product 7.6-inch diameter, 0.0020-inch radial OOR. Bottom: Crucible P/M product, 7.6-inch diameter, 0.0003-inch radial OOR. Radial magnification, 1000X.

Currently, ten of the most widely used high-speed tool-steel grades are being produced commercially by our P/M process. The field evaluation and application reports are substantiating the laboratory findings on the advantages of P/M high-speed steel. The process lends itself to the production of steel compositions heretofore unattainable by the conventional cast and hot-work production method. Thus, new upgraded high-speed steels producible only by P/M will be developed to take full advantage of this new process.

References

1. Kobrin, C.L., "Tool Steels Take the Powder Route," *Iron Age*, 200 (1967).
2. Dulis, E.J. and Neumeyer, T.A., "Particle-Metallurgy High Speed Tool Steel," ISI Publication 126, *Materials for Metal Cutting*, London: The Iron and Steel Institute (1967), 112.
3. Obrzut, J.J., "P/M Tool Steels Come Out Swinging," *Iron Age*, 207 (1971), 51.
4. Neumeyer, T.A. and Kasak, A., "Grain Size of High Speed Steels," to be published.

SESSION VI

HIGH-PERFORMANCE APPLICATIONS

MODERATOR: NORMAN B. SCHWARTZ
Senior Editor, IRON AGE

17. Potential Titanium Airframe Applications

R. H. WITT and O. PAUL

Grumman Aerospace Corporation
Bethpage, New York

ABSTRACT

Isostatically pressed and vacuum-sintered titanium alloy preforms from various sources were forged using conventional, high-energy rate and isothermal forging processes. Elemental powders were primarily employed to obtain the Ti–6Al–4V composition in the preforms. Results of evaluation studies pertaining to the effects of materials and process variables on microstructure, mechanical properties, and electron-beam weldability are presented. Selection of powder composition and forging process for production of airframe-quality titanium alloy forgings has been demonstrated. Potential airframe applications, producibility, and economic considerations are discussed, based on the results of the above evaluation.

Introduction

This chapter is concerned with investigating the feasibility of producing close-tolerance, aircraft titanium forgings from sintered preforms. Ti–6Al–4V titanium alloy preforms are to be produced by isostatically pressing elemental powders and sintering them in vacuum. If proven feasible, this process is expected to result in economic advantages by reducing machining costs and material scrap. In addition, the use of preforms has a definite economic potential in the development of forging parts using fewer operations and for producing shapes presently considered unforgeable.

The objective of the program is to determine physical and mechanical properties, dimensional characteristics, and weldability of Ti–6Al–4V titanium alloy forgings produced from isostatically pressed and sintered preforms. Initial design data for simple configurations (pancakes) have been obtained in the first part of the study, and more complex configurations representing actual aircraft components are being investigated. The component selected will satisfy the goal of the program, *i.e.*, to produce a deep-pocketed forging having uniform density and properties with draft angles no greater than two degrees (Figure 1).

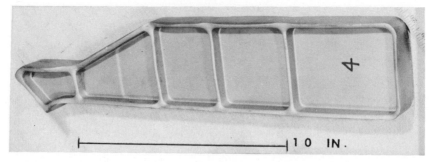

Figure 1. Deep-pocketed forging representing actual configuration of many aircraft components.

Background

The use of powder-metal preforms has been under investigation for a number of years for various alloys. The use of such preforms makes possible the following advantages in forgings:

1. Economical production of special preform shapes;
2. Production of parts to almost net dimensions (little or no machining required after forging);
3. Excellent machinability, where required;
4. Improved mechanical properties (finer grain size and uniformity of grain size).

In the case of titanium alloys, the potential savings are especially significant since titanium alloy forgings prepared by standard processes, which utilize wrought material preforms, must be purchased greatly oversize due to the inherent resistance of titanium to flow into die cavities. Thus, extensive machining is required.

An analysis of buy-weight to finished-weight of conventional currently produced titanium forgings shows that an average of three to eight pounds of raw forging must be purchased to obtain one pound of finish-machined part. Three to seven pounds per pound of finished forging (depending on forging size) must be machined away and consigned to the scrap pile. At approximately $6 per pound for titanium, the cost of this waste is intolerable; to this must be added the extra cost of machining away the excess material. A specific example of the potential savings for a typical deep-pocketed bulkhead part is presented in Table I.

Further savings can be accrued in the forging process when one considers the following conventional forging sequence starting from billets versus the close-tolerance P/M method starting with preforms.

TABLE I

Example of Savings Potential on Deep-Pocketed Part

Product	Mach 2 aircraft
Application	Bulkhead
Material	Ti–6Al–4V, annealed
Current process	Machined forgings
Cost, $/lb structure*	$136
New process	Forgings from sintered preforms
Cost savings, $/lb structure	$65
Raw forging weight (conventional)	232 lbs
Finish machined weight	39.8 lbs
Preform weight	50 lbs

*Based on 100 pieces.

Comparison of Conventional and P/M Forging Sequences

Conventional Forging Sequence	P/M Forging Sequence
Cut	Isostatic Press (rubber mold)
Upset (die #1)	Vacuum sinter
Block (die #2)	Forging (die #1)
Trim	Hand finish
Semifinish (die #3)	
Trim	
Finish forge (die #3)	
Warm trim	
Hand finish	

It is evident that die costs can be minimized, if techniques can be developed to obtain long die life. A number of trimming and heating operations are also eliminated.

Normally, conventional forgings require a 7° draft (angled sides) to permit removal of the forged part from the die. A low- or minimum-draft forging offers the potential for increased strength, since the strongest parts of the forging are the grains closest to the outer surface. Minimizing machining of forgings or, ideally, eliminating machining, leads to retaining the strength of the normal grain flow produced in the forging process. Briefly the advantages of low-draft forgings are as follows:

1. Finished part is stronger;
2. Savings in labor, materials, production time;
3. Production of close-to-net parts (draft and undercut minimum);
4. Wall and rib thickness can be held to minimum;
5. Machining costs lower;

6. Utilizing powder-metallurgy techniques, parts that were previously unforge-able can be economically produced;

7. The process is applicable to most shapes.

The last two advantages are made possible by use of the isostatic process, which allows the production of preforms to almost the size of the forged part. The fine grain size inherent in powders aids in the production of uniform strength, provided full densification is attained throughout the forged part. The use of elemental titanium and aluminum-vanadium master alloy powders in man-ufacturing Ti–6Al–4V titanium alloy preforms has significant advantages over prealloyed powders, due to higher green densities obtainable and the beneficial effects of the alloying process which occurs in the course of sintering. Both of these factors contribute to greater densification of preforms in the course of the sintering process. The as-sintered densities of preforms prepared from elemental powders can be as high as 95 percent of theoretical density.

One of the recent major evaluation programs [1] in the field of the titanium powder-metallurgical technology showed mixed results when prealloyed powders were employed. Results indicated that a 1750°F treatment to enhance prop-erties resulted in thermally induced porosity. The cost of prealloyed powders is also restrictive.

Preliminary Studies

The objective of these studies was to determine the feasibility of obtaining sound and 100-percent-dense simple configuration forgings with acceptable mechanical properties from sintered Ti–6Al–4V titanium alloy preforms manufac-tured from elemental powders. Three forging processes—high-energy rate, con-ventional, and isothermal—were evaluated in the course of this study. The con-figurations of the sintered preforms varied in accordance with the configurations of the forging dies available for use with each of these processes. All preforms were manufactured by isostatic pressing of elemental powders at 60 ksi and sin-tering of preforms at 2250°F in 10^{-4} mm (or better) vacuum. The character-istics of titanium powder utilized in this phase of study and some properties of sintered preforms are summarized in Tables II and III. This powder was repre-sentative of the commercial grades currently available in the range of $2–$4 per pound, and thus offered the greatest potential advantages from an economic standpoint. Some concern was expressed over the NaCl content of this powder (0.25 percent) and possible detrimental effects of this impurity level on the electron-beam weldability properties of forgings. The selection of this powder—in spite of the expressed reservations—was based on the data submitted by the selected vendor, which indicated that residual chloride can be removed by minor (proprietary) modifications of the sintering process. All preforms were delivered

TABLE II

Characteristics of Elemental Powders Utilized
in the Course of Preliminary Studies
Composition, Wt. %

Element	Elemental Ti Powder	Al–V Master Alloy	Sintered Preforms
C	0.01	N.D.	0.02
N	0.013	N.D.	0.039
O	0.10	0.07	0.18
H	0.004	n.d.	0.002
Na	0.099	n.d.	n.d.
Cl	0.15	n.d.	n.d.
Al	–	56.59	6.42
V	–	41.67	4.15

Note: n.d. = not determined.

Screen Analysis (typical)

Mesh	Percent
−100 + 200	42
−200 + 325	32
−325	26

TABLE III

Some Physical and Mechanical Properties
of Sintered Preforms

Property	Value
Density (% theoretical)	94.6–95.1
Tensile Strength, F_{tu}, ksi	114–122
Yield Strength, F_{ty}, ksi	97–104
Elongation	3–7 %
R.A. %	4.8–7.0

Metallography: No evidence of gross voids, inclusions, or inhomogeneity.
Radiography: Meets AMS 2635 film-density requirements.

in an as-sintered condition, and their density was in the range of 94–95 percent
of theoretical.

High-Energy Rate Forging Process (HERF)

The Dynapak High-Velocity Impact Forging process utilizes the principle of
quick-energy release to drive the ram at a speed in excess of 600 inches per sec-

ond. Since this is a balanced energy system, nearly all of the energy released is absorbed by the part being forged and the press does not require the massive support bases required in conventional forging. Since this process also appears to have a definite potential for yielding cost-effective methods for producing forgings with minimal die wear, and since "no draft" forgings were previously produced by this process, the HERF process was selected as the first one to be evaluated. The die utilized in this operation consisted of the lower block containing the cavity and the upper flat ram. The die assembly was preheated to about 400°F by manually operated torches prior to the operation. The preforms utilized in this phase of the study were rectangular shapes designed to yield $8'' \times 5'' \times \frac{1}{8}''$ and $8'' \times 5'' \times -\frac{1}{4}''$ configurations after 25-, 35- and 45-percent reduction on forging. Only one forging blow could be utilized in the course of the operation since, on separation of the die after the first blow, the specimen invariably "bounced" outside the die cavity. A significant temperature drop occurred on contact with cooler dies and, during the time required to reposition the preform inside the die cavity, this precluded the possibility of repeated blows. Prior to forging, the specimens were coated with Grafo 209 and Turco Pretreat coatings, dried in air, and preheated to 1750°F in an air furnace.

Preforms designed to yield $\frac{1}{8}$-inch-thick "pancakes" on forging failed to fill the die cavity. A complete fill on the die cavity was obtained only on forging of thicker preforms designed to yield $\frac{1}{4}$-inch-thick pancakes. Figure 2 shows the microstructure of the preform in the as-sintered condition. Figures 3 and 4 show the microstructure of forgings after 25-percent and 45-percent reduction in thickness, respectively. From Figures 3 and 4, it is evident that, although a significant densification of preforms could be obtained on forging, microporosity remained

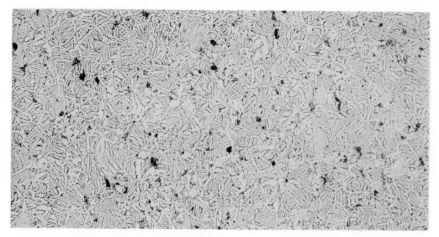

Figure 2. Microstructure of a preform in an as-sintered condition. Etch: modified Keller's soln. 100X.

Figure 3. Microstructure of a forging after 25% reduction by HERF process. Etch: modified Keller's soln. 100X.

Figure 4. Microstructure of a forging after 45% reduction by HERF process. Etch: modified Keller's soln. 100X.

at the grain boundaries even after 45-percent reduction in thickness. Tensile specimens were machined from $\frac{1}{4}$-inch-thick pancakes reduced 25–45 percent in thickness by forging, coated with Turco Pretreat and Formkote T-50 coatings (mixture used in most in-house stress-relieving operations on titanium alloys) and stress-relieved at 1300°F for two hours. The results of the tensile tests revealed that although the tensile and yield strength values obtained exceeded the requirements of Mil-T-9047, the elongation values obtained were in the range of 2–4 percent and were thus far below the minimum requirements specified (10 percent). Based on the results of metallographic and tensile evaluations, it was evident that the operating temperature of the forging dies and mass of preforms will have to be increased to improve densification of preforms on forging. Both of these goals were achievable, utilizing an existing forging setup previously used in manufacturing of experimental parts by conventional forging techniques.

Conventional Forging Process

Forgings produced by this process from wrought and powder-metallurgical preforms are shown in Figure 5. The temperature of the forging die was maintained at 800°F. No difficulties were encountered in filling the die cavity by this method. Metallographic examination revealed that although the extent of densification was considerably improved, compared to the previously studied (HERF) process, some isolated porosity was still detectable in the forging at 100X magnification, as shown in Figure 6. The sample shown was taken from the center section of the forging; more extensive porosity was detected in the rib section. The specimens for tensile tests were machined from the flat portion of the forging and stress-relieved at 1300°F for two hours. The tensile and yield strengths

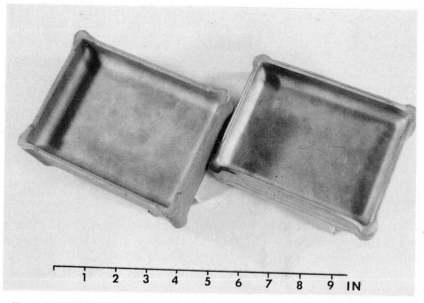

Figure 5. Ti–6A1–4V forgings produced from wrought stock (left) and powder preforms (right) by conventional forging techniques.

Figure 6. Microstructure of conventionally forged preforms.   Etch: modified Keller's soln. (95 H_2O:2.5 HNO_3:1HF). 100X.

of the forgings exceeded the applicable requirements of Mil-T-9047. Elongation, however, was in the range of 3-5 percent, and was thus below the required minimum (10 percent).

Isothermal Process

The next logical step was to investigate the isothermal approach, where the object to be forged and the forging die are at the same temperature. The flow of material is thus enhanced and the heat losses are practically eliminated.

A forging produced in the isothermal die is shown in Figure 7. The powder preforms used in this study were rectangular blocks ($1\text{-}\frac{1}{4}'' \times 1\text{-}\frac{1}{4}'' \times 7\text{-}\frac{1}{2}''$). The die was preheated to 1750°F in a separate furnace, and in the course of the operation its temperature was maintained in the range of 1650–1750°F by torches positioned around the die assembly. All specimens were forged at a pressure of 30 ksi. The metallographic studies established that forgings produced by the isothermal method were completely free from porosity; metallography, however,

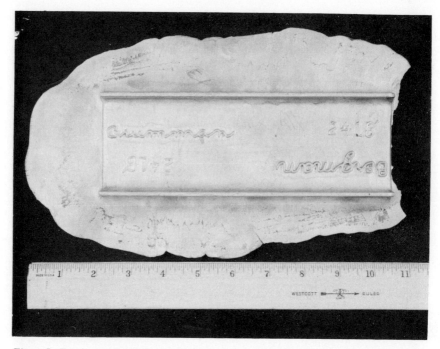

Figure 7. Forging produced by the isothermal process in the course of preliminary studies.

revealed extensive inhomogeneity regions, as shown in Figure 8. The tensile specimens were machined from the flat portion of the forging and stress-relieved at 1300°F for two hours. The results of the tensile tests established that tensile and yield strengths exceeded the requirements of Mil–T–9047; the elongation, however, was in the range of 5 percent, and thus did not meet the minimum requirements specified (10 percent). Electron-beam welding of forged and machined components was accompanied by excessive sputtering ("Roman Candle" effect) and arcing, and resulted in excessive porosity in the welds. Since the test results strongly indicated the presence of chloride impurities, a chemical analysis was conducted on forged specimens. The obtained data established that 0.17 percent of 0.25% NaCl reported in the original powder remained in the forging after all phases of processing.

Powder-Selection Studies

Results of preliminary studies conducted led to the following conclusions:
1. 100-percent-dense forgings are obtainable from P/M preforms by the commercial (and relatively simple) isothermal forging process.

Figure 8. Microstructural inhomogeneity detected in isothermal forgings produced in the course of preliminary studies. Etch: modified Keller's soln. 100X.

2. Isothermally forged preforms met tensile and yield requirements specified in Mil-T-9047. Problems were encountered in meeting elongation and electron-beam weldability requirements. These problems appeared to be associated with the impurities contained in the elemental titanium powder utilized in manufacturing and, specifically, to the presence of NaCl, although it has been reported that NaCl can be removed in the course of sintering at $2250°F$ in 10^{-4} torr vacuum. Our investigations established that a substantial portion (0.17 percent) of the original chloride remained in the material after all phases of processing. It is evident on the basis of preliminary studies that titanium powder produced from sponge without an intermediate purification process cannot be utilized in production of aircraft-quality forgings, even though they are very attractive from an economic standpoint. Accordingly, further studies were limited to purified grades of elemental titanium powder. These powders, although somewhat more expensive compared to unpurified grades, still offer an attractive economic potential compared to conventional (wrought) materials or prealloyed grades. A plan of action was developed in accordance with which small quantities of preforms were to be prepared by various vendors, utilizing purified grades of titanium powder and Al-V master alloy and processed to evaluate the weldability, tensile, and microstructural characteristics of resulting forgings.

TABLE IV

Preform Variables—Isothermal Forgeability Study

	Vendor					
	A		B		C	D
Powder type	E + MA	PreA	E + MA	E + MA	E + MA	E + MA
Ti powder mfg.	H/D	H/D	Prop.	Exp. Prop.	CR + P (Prop.)	Mg–red
Compaction	Iso.	Iso.	Iso.	Iso.	Iso.	Mech.
Composition	Std.	Std.	ELI	Std.	ELI	Std.
No. of preforms	4	2	6	1	6	6
Density, % theor.	98 min	<90	95 min	98.5 min	94 min	n.d.

Explanation of symbols:

E + MA = Elemental titanium and Al–V master alloy.
PreA = prealloyed Ti–6Al–4V powder.
H/D = Hydride/dehydride.
Prop. = Proprietary.
Exp. = Experimental.
CR + P = Chemical reduction and purification.
Iso. = Isopressed.
n.d. = Not determined.

The data on characteristics of powders and properties of sintered preforms prepared by four different vendors are summarized in Table IV. Three of the vendors utilized purified grades of titanium powder in manufacturing of preforms; the fourth vendor (identified as vendor D in the table) used a standard (unpurified) grade of titanium powder prepared by magnesium reduction of the sponge rather than sodium reduction used predominantly throughout the industry. These preforms were included in the present study to evaluate possible effects of the different reduction processes on properties of forgings. Specimens were removed from each group of the preforms to evaluate the electron-beam weldability characteristics of preforms in an as-sintered condition. Sound welds could be produced only in preforms produced by vendor A, as shown in Figure 9. The electron-beam welds produced using identical parameters in preforms prepared by other vendors contained excessive porosity (Figure 9).

A series of experiments was also conducted to evaluate three commercially available coatings for potential application in forging of titanium powder preforms. The utilization of commercially available coatings, rather than those of proprietary composition used by many forging vendors, appeared to be advisable

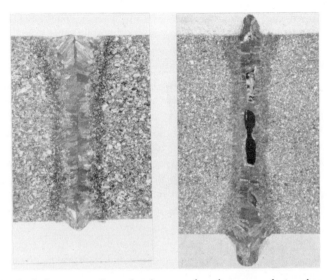

Figure 9. Typical cross sections showing sound and porous electron-beam welds in sintered preforms.

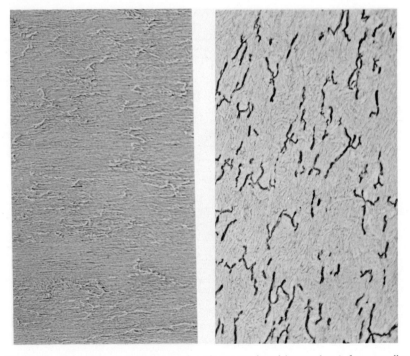

Figure 10. As-forged microstructure of preforms produced by vendor A from preallyed Ti–6Al–4V powder. Left, as-etched polished; right, as-etched. 100X.

in order to ascertain that all parameters of the forging process could be dupli-
cated by many forging vendors involved in the program. The experiments consisted of heating coated specimens of as-sintered preforms at 1750° in an air
furnace and subsequently evaluating the microstructure for the presence and
thickness of the α-layer on the surface. The best results were obtained with
Markal CRT coating for which practically no α-layer was detected on the surface.
Prior to the isothermal forging, one specimen from each group of the preforms
was coated with Markal CRT compound; the remaining preforms were coated
with proprietary coatings normally utilized by the vendor in forging of titanium
alloys.   There were no significant variations in the surface characteristics of
forgings precoated by both methods.

Metallographic examination revealed that complete densification of preforms
could be obtained only in Vendor A preforms prepared from both elemental and
prealloyed powders. (Note: several preforms were prepared by Vendor A from
prealloyed powders to permit a comparison of the results of the present study
with those of a major evaluation program conducted utilizing prealloyed grades).
The microstructures of forgings prepared from both types of Vendor A preforms
are shown in Figures 10 and 11. Figure 12 shows the microstructures of forgings

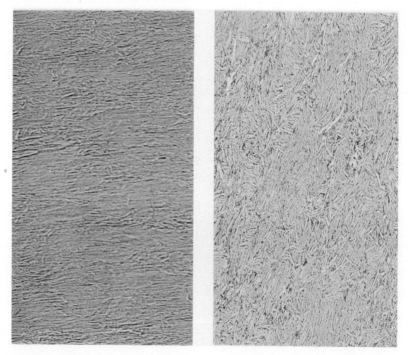

Figure 11. As-forged microstructure of preforms produced by vendor A from elemental
Ti powder and A1–V master alloy. Left, as-etched polished; right, as-etched. 100X.

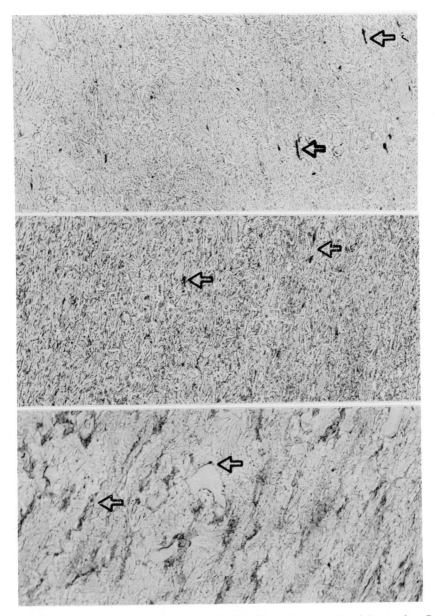

Figure 12. As-forged microstructure in Ti–6A1–4V preforms produced by vendors B (top), C (center), and D (bottom). Arrows indicate sites of remaining porosity. Etch: modified Keller's soln. 100X.

produced from preforms prepared by the other three vendors; it is evident that a complete densification could not be achieved on isothermal forging of these preforms. The forgings prepared from the standard grade of titanium powder (magnesium-reduction process) exhibited a highly nonuniform grain structure and indicated extensive regions of segregation and/or contamination which were very similar to those detected previously in the case of the standard grade of titanium powder prepared by the sodium-reduction process. Specimens for tensile tests were machined from the flat portion of the forging and stress-relieved at 1300°F for two hours.

TABLE V

Properties–Isothermal Forgeability Study

	Vendor					
	A		B		C	D
Powder type	E + MA	PreA	E + MA	E + MA	E + MA	E + MA
Forged density % theor.	100	100 (1)	(3)	(3)	(3)	(3) (2)
F_{tu}, ksi	152.4	71.2		140.4	145.3	125.6
	150.2	82.3		140.0	144.8	115.0
F_{ty}, ksi	151.7	(4)		138.4	138.2	124
	148.0	(4)		139.0	138.2	(4)
Elong. % (1″ gage length)	14	Nil		9	7	1
	13	1		8	8	Nil

Mil-T-9047 requirements: F_{tu} = 130 ksi minimum.
$\quad\quad\quad\quad\quad\quad\quad\quad\quad$ F_{ty} = 120 ksi minimum.
$\quad\quad\quad\quad\quad\quad\quad\quad\quad$ Elong. = 10% minimum.

Notes: (1) Intergranular phase detected.
$\quad\quad\quad$ (2) Structure inhomogeneity.
$\quad\quad\quad$ (3) Porosity.
$\quad\quad\quad$ (4) Premature failure.

The test results summarized in Table V show that tensile, yield, and elongation properties of forgings manufactured from Vendor A *elemental* powders exceeded the minimum requirements specified by Mil-T-9047. Accordingly, the experimental data obtained establishes the commercial availability of powder which can yield 100-percent-dense forgings with acceptable tensile, yield, and elongation characteristics. This finding constitutes a successful conclusion of the feasibility study.

Current Status

Preforms are currently being prepared to be isothermally forged to the complex configuration shown in Figure 1. Following this, tensile, fatigue, and stress-corrosion characteristics will be established.

Summary

The investigation has shown that sound, high-strength, ductile Ti-6A1-4V titanium forgings can be produced from isostatically pressed elemental powder preforms using the isothermal forging process. These properties are attainable, utilizing the standard postforging treatment of 1300°F for two hours. It remains to demonstrate that complex forgings of high integrity can be forged having acceptable fatigue and stress-corrosion properties for airframe applications.

Acknowledgment

This work is being supported by the Naval Air Systems Command, Washington, D.C. [2].

References

1. Peebles, R.E., "Titanium Powder Metallurgy Forging," General Electric Company, Cincinnati, Ohio, Air Force Materials Laboratory Contract Report AFML-TR-71-148, Final Report, September 1971.
2. Witt, R.H. and Paul, O., "Feasibility of Producing Close Tolerance F-14 Titanium Forgings in Sintered Preforms," Grumman Aerospace Corporation, Bethpage, New York, Naval Air Systems Contract Report, to be published.

18. Processing of High-Performance Alloys by Powder Metallurgy

R. M. PELLOUX
Massachusetts Institute of Technology
Cambridge, Massachusetts

ABSTRACT

Processing of high-performance alloys by powder metallurgy offers a means to eliminate segregation of alloy constitutents and to minimize the size of the inclusions. Powder-metallurgy processing of oxide dispersion-strengthened alloys, nickel-base superalloys, and maraging steels is discussed and the mechanical properties of the P/M alloys are compared to the properties of the cast and wrought alloys.

Introduction

For the purpose of this chapter, we shall define high-performance alloys as alloys having a yield strength–elastic modulus ratio above 5×10^{-3} and usually close to 10^{-2}. The strengthening of these high-performance alloys is obtained through the uniform distribution of very fine precipitates (less than 500 Angstroms in size) in a solid-solution matrix. Since the alloy chemistry is quite complex, it is difficult to obtain a homogeneous wrought alloy by the ingot–hot work process. The best ingot solidification control cannot avoid macro- and microsegregation of some of the alloying elements. The original ingot segregation is retained in the finish part in spite of the usually long homogenerization times. A typical example of segregation is the banded structure of maraging steels attributed to a nonuniform distribution of molybdenum. The iron-silicon-rich clusters of inclusions and constituent particles in high-strength aluminum alloys are another example of this processing problem. In all these alloys, inclusions or defects range in size from 1–25 microns—that is, three orders of magnitude larger than the required strengthening precipitates.

Processing research during the last twenty years has shown that even if one were to start with raw materials of the highest purity, the presence of inclusions or of undissolved constituent particles in the finished product cannot be avoided.

351

For many alloy systems, careful processing has reduced the volume fraction of inclusions to 0.1 percent, and even to 0.01 percent. However, the remaining inclusions, which still range in sizes from 1–25 microns, have a strong influence on tensile ductility, fracture toughness, and fatigue strength. In fatigue, for instance, it is well known that crack initiation is controlled by the size of the inclusion rather than by the volume fraction of inclusions—one large inclusion is as detrimental as many.

Consequently, it is clear that the ingot–hot work processing route will never allow a complete control of the microstructure of a complex alloy. By contrast, the P/M approach provides an excellent way to refine the microstructure of alloys right from the beginning of the manufacturing process.

This presentation is a review of some recent research work on the preparation of high-performance alloys by powder metallurgy. The different alloy systems which will be reviewed include: (1) oxide dispersion-strengthened alloys; (2) nickel-base superalloys; (3) maraging steels.

Oxide Dispersion-Strengthened Alloys

The high-temperature strength of oxide dispersion-strengthened alloys is due to the interaction between a stable dislocation network and a uniform distribution of oxide particles less than 500 Angstroms in size. The volume fraction of oxide particles required is on the order of 1–3 volume percent. The best way to achieve this uniform distribution of oxide precipitates is to oxidize one of the solute elements of a master alloy. For instance, Grewal [1] prepared two nickel-beryllia alloys containing, respectively, 1 and 2 volume percent BeO by oxidizing the beryllium solute of nickel alloys containing 0.15 and 0.32 weight percent beryllium. The oxidation of the solute element is accomplished by internal oxidation or surface oxidation of fine alloy powders.

The manufacturing process starts with alloy particles (the word "particles" refers to any size or shape of alloy powders). These particles can be machining chips from a master ingot or inert gas-atomized powders. The machining chips are ball-milled to particles less than 150 microns. The atomized powders have an average diameter of 25–30 microns. Both types of particles are mechanically reduced to flakes (size 0.1×100 microns) in an attritor. Figure 1 is a scanning electron micrograph of the flakes after attrition. In general, surface oxidation of the attrited flakes is sufficient to oxidize all the solute element. Complete oxidation can also be obtained by internal oxidation in a stream of hydrogen and water vapor with a H_2O/H_2 ratio on the order of 0.3. An exposure time of 10 hours at $950°C$ is sufficient for complete internal oxidation of the nickel-beryllium alloys. In order to have complete diffusion of oxygen through a particle in a relatively short diffusion time, the particles should be less than 5 mi-

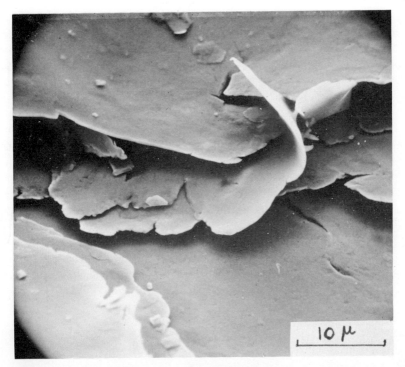

Figure 1. Flake of a nickel 0.15 wt. % beryllium following attrition and prior to surface oxidation.

crons thick—otherwise an oxide skeleton is formed at the surface, leading to a nonuniform distribution of the oxide particle sizes. The final step of the process consists in hydrogen reduction of the less stable oxides formed during attrition and internal oxidation.

Compaction of the O.D. powders id done by cold pressing, vacuum sealing in a can, and hot extrusion. The extrusion ratio and temperature are such that adiabatic heating during extrusion will not raise the billet temperature above the coarsening temperature of the oxide particles. Following extrusion, the alloys are hot or cold worked to obtain the desired mechanical properties. In Ni–BeO alloys processed by surface oxidation, the BeO particle size is around 250–300 Angstroms. These alloys with a 100-hour rupture life under 10,000 psi at 1800°F are as strong as TD nickel, but they are easier and cheaper to process. The beryllia coarsening rate is lower than for thoria, and the average recrystallization temperature is above 2200°F.

A nickel-base alloy containing 4.5 percent molybdenum and 2 volume percent BeO was processed by the same technique without any difficulty. It should also be mentioned that all the alloys processed by surface and internal oxidation which do contain 0.1–0.2 volume percent of oxide particles are the result of the nonuniform oxidation of large alloy particles. However, the uniform distribution of these larger particles does not appear to be detrimental to the tensile and creep properties of the alloys.

Nickel-Base Superalloys

Most of the nickel-base superalloys are cast since their hot plasticity is very limited. Although a cast microstructure is excellent for creep resistance, it has poor fatigue strength since the large carbides and the primary γ' particles serve as fatigue-crack initiation sites. On the contrary, a uniform and refined microstructure can be produced by powder metallurgy. The alloy selected for this program was IN-100 which has the following composition: 10Cr, 15Mo, 4.7Fe, 5.5Ti, 1V, balance nickel.

Argon-atomized powders were produced by Federal Mogul from a prealloyed vacuum-cast ingot. The average particle size is less than 150 microns, and Figure 2 is a typical illustration of the shape and size distribution of the F-M powders.

Figure 2. Argon-atomized IN-100 powder.

Coarse powders (size on the order of 500 microns) produced by the spinning electrode process were supplied by Nuclear Metals. The coarse N-M powders are shown on Figure 3. Cross sections of the two different powders are given in Figures 4 and 5. Each powder particle is sound and free from internal porosity or inclusion. The average dendrite arm spacing is on the order of 5×10^{-4} to 10^{-4} inches. The dendrite network is characteristic of the process used in quenching the atomized powders. Since one of the most critical requirements for a successful P/M process is the cleanliness of the powders, the surfaces of the powder particles were evaluated by scanning electron microscopy and by non-dispersive X-ray analysis. The surfaces were found to be free of oxide and carbide contamination. The oxygen content in the consolidated alloys range from 50-100 ppm.

The powders were compacted to fully dense billets by two different processes following canning in steel cans and vacuum degassing for 2 hours at 900°F to a vacuum better than 5×10^{-5} torr. Hot isostatic compaction at 2320°F and 25,000 psi for one hour was used for the F-M powder. The billet was fully dense but each particle was surrounded by a network of titanium carbide, which is shown in Figure 6. The examination of the as-received powders gave no indication of the presence of this phase at the surface of the particles, and its forma-

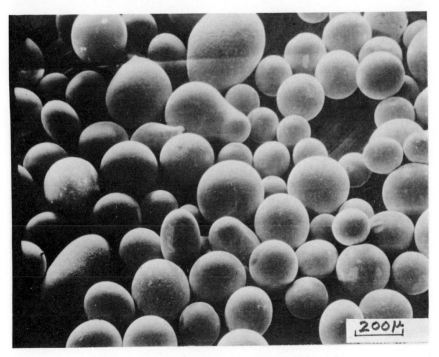

Figure 3. Rotating-electrode-atomized IN-100 powder.

Figure 4. Section of inert gas-atomized IN-100 showing dendrite arm spacing.

tion was the result of the HIP cycle. We found recently that the TiC phase can be completely eliminated by using a master alloy of IN-100 with a low carbon content.   The carbide network limited the high-temperature ductility of the alloy, which failed by decohesion along the particle boundaries during slow and fast strain-rate tests.

Compaction of the two types of powders by hot extrusion at 2150°F was very successful and produced a uniform microstructure with a grain size on the order of 10 microns.   The average size of the second-phase particles (carbides and primary γ') does not exceed 2 microns.

The room temperature mechanical properties following extrusion are given in Table I. The small grain size of the extruded material accounts for the large increase in yield strength over the cast material.   The improvement in ultimate tensile strength is due to the elimination of the large brittle second-phase particles which allows a high ductility at fracture. These excellent room-temperature tensile properties are retained at elevated temperatures (up to 1400°F) as long as the test temperature is below the equicohesive temperature. At 1800°F, the creep strength is strongly dependent upon the grain size. Table II gives the rupture life at 1800°F for a stress of 10,000 psi.

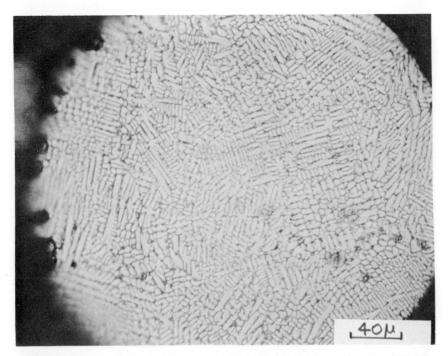

Figure 5. Section of rotating-electrode-atomized IN-100 showing uniformity of dendrite arm spacing.

The grain-coarsening heat treatment has produced a uniform grain size of 150 microns, which is still an order of magnitude smaller than the cast grain size. It is hoped that grain coarsening should be easier with a low-carbon IN-100. Although the creep properties are not as good as for the cast alloy, it is expected that the high-temperature fatigue behavior of the extruded alloy will be far superior because of the absence of brittle phases.

The total elongation obtained during creep rupture at 1800°F range from 40 percent for 1 hour life to 200 percent for a life of 100 hours. These results indicate that the fine grain size IN-100 was superplastic at 1800°F. A series of low strain-rate tensile tests between 1700°F and 2100°F confirmed the super-plastic behavior of IN-100 above 1800°F at strain rates below 0.02/min. The superplastic coefficient in the equation $\sigma = A\epsilon^m$ is on the order of 0.5. These results are in agreement with the work of Reichman [2], who has also reported the superplastic behavior of powder-processed alloys of IN-100.

Superplasticity in IN-100 can be used to form parts which can subsequently be heat-treated to coarsen the grain size and improve the creep strength above 1400°F. Our current research effort is directed towards the understanding of the mechanisms of superplasticity in these alloys.

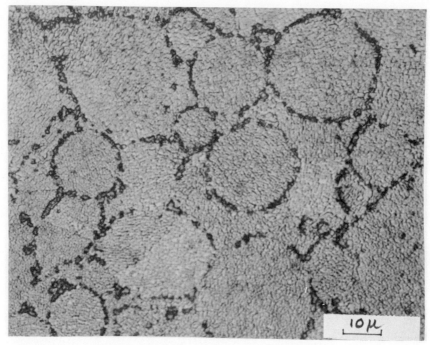

Figure 6. Microstructure of hot isostatically pressed IN-100 powder produced by inert gas-atomization. Each particle is surrounded by a TiC network.

Maraging Steel

A high-strength maraging steel (VM 300)(18Ni-9Co-5Mo-0.6Ti-0.1Al) was selected as a high-performance alloy to be processed by powder metallurgy. Most maraging steels show some degree of banding due to the segregation of alloy constituents (titanium and molybdenum) in the original ingot. It was expected that billets produced from atomized powders should be free from banding and inclusion stringers.

Two different sources of atomized maraging powders were used: (a) Steam-

TABLE I

Room-Temperature Mechanical Properties

	Yield Strength (ksi)	Ultimate Tensile (ksi)	Elongation (%)
IN-100	123	147	9
Extruded F-M powder	175	244	20
Extruded N-M powder	171	238	21

TABLE II

Stress-Rupture Strength at 1800°F and at 10,000 psi

	Rupture Life
IN-100 as-cast	
grain size over 3,000 microns	4,000 hrs
IN-100 extruded	
overaged 24 hrs, 1800°F	
grain size = 10 microns	4 hrs
IN-100 extruded	
24 hrs, 2270°F for grain coarsening	
20 hrs, 1825°F overaging	
grain size = 150 microns	300 hrs

atomized powders—steam atomization is cheap and it gives a high quenching rate. The surface oxide layer produced by the steam can be removed by a proper pickling treatment. Steam-atomized powders contained, however, a large number of small oxide inclusions rich in titanium and aluminum. There was a corresponding loss of titanium and aluminum in the matrix of the alloy. (b) Rotating-electrode powders—these powders are very clean with a dendrite arm spacing on the order of 10 microns. Figure 7 shows the typical shape and size of rotating-electrode powders. The dendrite structure is readily visible on the outside surface of the particles at high magnification (Figure 8).

The two types of powders were compacted either by hot extrusion or by hot isostatic compaction, or by both processes combined—that is, HIP followed by extrusion. A standard heat treatment was given to all the alloys: 1 hour solution at 1500°F air cooled; 3 hours aging at 900°F. The room-temperature mechanical properties are given in Table III.

The steam-atomized and extruded alloys show a large number of small oxide inclusions which account for the low reduction in area. All specimens show a reasonable strength level in the aged condition, considering the loss of titanium and aluminum. The strength and ductility of the rotating-electrode powders compare favorably with the commercial grade of maraging steel. The differences between hot isostatically pressed plus extruded and direct powder-extruded material are minimal. Since the room-temperature tensile properties of the P/M alloys compare very well with the wrought alloys, it was decided to investigate the fatigue and fracture-toughness properties of the P/M alloys.

S-N curves were determined for the rotating-electrode P/M alloy (HIP and extruded) and for the commercial grade. The fatigue strength at 10^6 cycles is on the order of 70 ksi for the two alloys, which is a low value but typical of the poor fatigue-endurance strength of the maraging steels. The reduction in inclusion sizes and volume fraction of the P/M alloy does not improve the fatigue

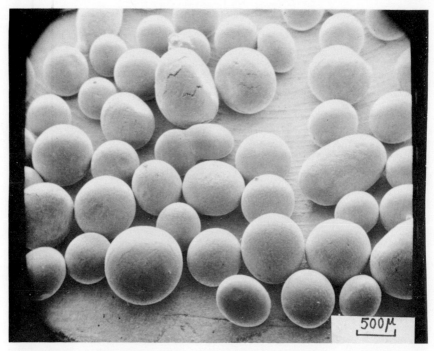

Figure 7. Maraging steel powder atomized by the rotating-electrode process.

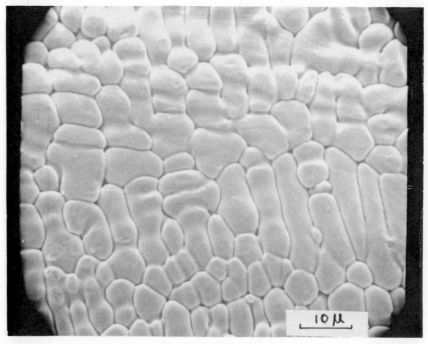

Figure 8. Surface of the maraging steel powder of Figure 7, showing the average dendrite spacing.

TABLE III

Mechanical Properties of Maraging Steels

	0.2% Yield Strength (ksi)	UTS (ksi)	Reduction in Area
Commercial Alloy			
Solutionized	138	145	76
Aged	269	274	50
Powder Processed			
Steam-atomized, extruded			
Solutionized	100	163	46
Aged	235	247	12
Rotating electrode, extruded			
Solutionized	125	153	68
Aged	285	290	48
Rotating electrode, HIP			
Solutionized	123	147	22
Aged	−	−	−
Rotating electrode, HIP + extruded			
Solutionized	125	151	65
Aged	285	291	47

strength over that of the commercial alloy. Some recent work has shown that the low fatigue strength of maraging steels is due to an environment effect, and a systematic comparison of the two alloys should be done in an inert environment such as dry argon.

Fracture toughness K_{1C} was also measured for the commercial grade and the P/M grade of maraging 300. The P/M billet was produced by HIP compaction of rotating-electrode powders followed by a 4/1 reduction by rolling. Figure 9 illustrates the results. The P/M alloy shows extensive microdelamination along the fracture surface. These delaminations correspond to debonding along particle boundaries under the transverse stresses present near the crack tip in a plane-strain condition. The fracture-toughness values are given in Table IV.

The fracture toughness of the P/M alloy is good, although it is lower than for the commercial alloy; however, the appearance of the fracture surfaces indicate that perfect bonding of the powder particles is difficult to achieve even after a 4/1 hot rolling reduction.

Conclusions

The processing of high-performance alloys by powder metallurgy offers a means to eliminate segregation of alloy constituents and to minimize the size of

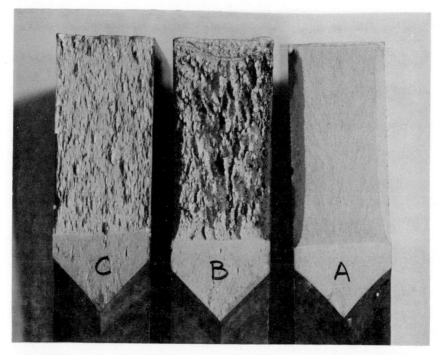

Figure 9. Fracture surfaces of fracture-toughness test bars of maraging steel. (A) Commercial grade K_{1C}, longitudinal = 63 ksi$\sqrt{\text{in}}$. (B) Powder-processed K_{1C}, transverse = 63 ksi$\sqrt{\text{in}}$. (C) Powder-processed K_{1C}, longitudinal = 51 ksi$\sqrt{\text{in}}$.

the inclusions. This refinement of the microstructure is achieved by the high quenching rates obtained during atomization. Compaction of coarse powder particles (sizes up to 500 microns) by hot isostatic pressure or by hot extrusion produces fully dense billets which are comparable to the wrought products but with superior homogeneity. Powder-metallurgy processing of nickel-base super-alloys has a great potential since the improvements in workability and mechanical properties of the P/M alloys over the cast alloys are outstanding.

TABLE IV

Fracture Toughness of Maraging Steel
(ksi$\sqrt{\text{in}}$)

	Longitudinal	Transverse
Commercial	63	71
P/M alloy	51	63

Acknowledgments

This work was supported by the Advanced Research Project Agency. The contributions of M. Grewal, L. Moskowitz, and L. Van Swam are gratefully acknowledged.

References

1. Grenwal, M.S., "Dispersion Strengthened Nickel Base Alloys Produced by Internal and Surface Oxidation," Unpublished Sc.D. dissertation, Massachusetts Institute of Technology, February 1972.
2. Reichman, S.H. and Smythe, J.W., "Superplasticity in P/M IN100 Alloy," *Int. J. Powder Met.*, 6 (1970), 65.

19. Production System and High-Performance Automotive Applications

R. F. HALTER

Cincinnati Incorporated
Cincinnati, Ohio

ABSTRACT

The hot forging of P/M parts for automotive applications has received increased attention during the past year. Numerous investigators have developed many parts for high-performance applications. These parts range from differential pinion gears to transmission ring gears and to connecting rods for the engine and air-conditioning compressors.

The properties, performance, and economics of these parts has in most cases equaled or surpassed those of the conventionally wrought and machined parts. However, hot forging of P/M parts has brought about new problems and process limitations which are discussed.

Introduction

The advantages of the P/M approach to forging have been observed and reported by many investigators. Advantages such as more homogeneous structure, fine grain size, savings in materials, improved mechanical properties, and reduced production cost have been cited. For these reasons, the automotive industry is inclining more and more toward the P/M forging approach. However, they must establish the parameters of property control, dimensional control, and economics for each specific application.

The successful achievement of the desired parameters, namely, impact strength, hardenability, and economics has been demonstrated on a differential side pinion gear [1,2]. This high-performance application and others will be discussed in this chapter.

Process

General Description

The powder-metal forging process is shown diagramatically in Figure 1. The first step in the process is the selection of the alloy powder. In almost all instances, the selection of the alloy was based on use of the present alloy to produce the part from wrought stock. However, with the advantages of the P/M approach to forging, the alloy selection should be based on economics and required final-part properties.

The second step is the compaction of the preform. This step is usually carried out in the conventional manner with the use of admixed lubricants. However, as the size of powder-metal preforms increases, the need for die-wall lubrication may become quite important.

The third step is the sintering of the preform. Again, the sintering operation has usually been carried out in a conventional manner for most ferrous preforms. Some high-temperature sintering has been carried out in order to achieve greater ductility during forging. The specifications of sintering will depend on the forging requirements and the desired final-part properties.

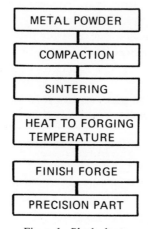

Figure 1. Block chart.

The fourth step is the heating of the preform to the desired forging temperature. This step could be combined with the sintering operation by providing a short cooling zone on the furnace to bring the preforms down to the desired forging temperature. This type of system could present problems, especially when the forging press is shut down for a tooling change. Greater versatility, control, and higher production rates can be achieved with the use of induction heating.

The fifth step is the finish-forge operation. The production of a precision forging without flash usually requires one blow of the forging press. If dimensional tolerances less than ±0.002 or ±0.003 are required, then finish-machining operation may be necessary.

Process Variables

In order to optimize the process, considerable effort has been spent on evaluating the effects of the process variables on the final-part properties. It has been found that preform shape, preform weight control, the use of atmosphere in the induction coil, preform temperature, tooling temperature, and lubrication had significant effects on the tooling life and the final-part properties.

Some of these variables have been discussed in a previous chapter and the others will be discussed later in this chapter in conjunction with the development of different parts [2].

Applications

Differential Pinion Gear

The side pinion geat was attractive to the automotive industry for P/M forging because it offered: [1]
1. High potential volume;
2. High strength requirements;
3. Elimination of excessive material loss;
4. Uniform shape and size;
5. Tooling cost that was not excessive.

Three different pinions have been developed by Cincinnati Incorporated. The alloys used in these three feasibility programs were AISI 4620, 4023, and 8627.

The first development program was on the AISI 4620 pinion (Figures 2 and 3). The program was a cooperative effort between Delco Moraine Division of General Motors and Cincinnati Incorporated. Several problems were encountered and considerable knowledge was generated in the development of this gear.

The first problem was the determination of the best preform shape. Tests were run using plasticine (modeling clay) preforms in order to determine the best flow pattern. Based on this study, the partial-bevel preform was selected [2].

The second problem was that weight control of ±0.5 percent was required to hold the height dimension within ±0.003. This was accomplished by use of a tonnage-control system on the compacting press.

The first run of several thousand gears was found not to have sufficient impact strength. Examination of the fractured surface of a gear with a scanning electron microscope showed that the structure was nonhomogeneous and contained numerous microcracks. The densification and metal flow during the forging process sheared the initial particle bonds, and many of the surfaces were never healed under the compressive loads at the bottom of the forging stroke. The

Figure 2. Differential side pinion gear and preform.

Figure 3. Positional relationship of the two side gears (large gears) and the two pinion gears (small gears) in a differential.

balance between the die temperature and preform temperature was increased 100°F, and this solved the problem.

During the initial forging runs to determine the exact weight of the preform, excessive porosity was found on surfaces which were not constrained during the forging operation. Other investigators have found this same problem [3]. In fact, fracture of the nonconstrained surfaces occurred if the part did not fill the entire die cavity. The preform weight had to be sufficient to insure that the entire part was under compression at the bottom of the forging stroke.

Another area of concern for most investigators was that of oxidation during the heat-up for forging and during the transfer to the forging dies. It has been shown that an uncoated preform can be exposed to an oxidizing atmosphere for short periods without any noticeable oxidation or decarburization [4]. This time period is sufficient to allow the part to be transferred to the die without protection. However, it is not sufficient in most cases to permit the preform to be heated to the desired forging temperature. The introduction of a non-oxidizing atmosphere into the induction coil, coating the preform with a graphite lubricant, or both, will be necessary to protect the preform. By coating the pinion preforms with a graphite lubricant prior to their introduction into the forging system, additional atmosphere protection was found to be unnecessary.

This summarizes several of the problems encountered in the development of the P/M forging process for the differential pinion gear.

The results of the program showed that the process is repeatable and controllable. Many thousands of parts have been forged with excellent dimensional reproducibility and superior mechanical properties.

Gears from this pilot production system were put in a test program consisting of a variety of physical tests, metallurgical tests, and car tests.

The physical tests included hardenability and impact-strength tests on finished gears. The hardenability and impact strengths of the P/M gears was equal to and greater than the values obtained on the wrought gears [1].

The metallurgical tests were concerned with the evaluation of the microstructure. The ASTM grain-size number of the P/M gears was 11, and the ASTM grain-size number of the wrought gear was 7, as shown in Figures 4 and 5. Despite this small grain size, no loss in hardenability was observed [1,5]. The structure of the gear was relatively free of impurities, as shown in Figures 6 and 7. The density in this area is approximately 99 percent of theoretical.

The car tests included mileage tests and supervised G.M. proving-ground tests on several cars. One test was durability mileage, a scheduled 36,000-mile G.M. proving-grounds test. This is an accelerated abuse test which programs the car through various stops, starts, speeds, and road conditions.

Another was the 10,000-mile durability hill schedule; a proving-ground test that severely loads drive components and transmissions. This test consists of accelerated hill climbs and sharp turns as part of the over-all schedule at the G.M. proving grounds.

Figure 4. Grain-size comparison of P/M forged pinion gear. Bulk tooth structure. Nital etch. Not heat-treated. 100x.

Figure 5. Grain-size comparison of wrought pinion gear. Bulk tooth section. Nital etch. Not heat-treated. 100x.

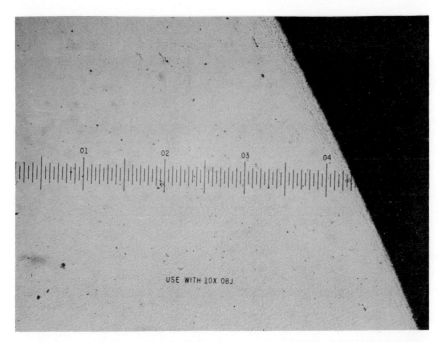

Figure 6. As-forged condition of P/M pinion gear on pitch line. Unetched. 100x.

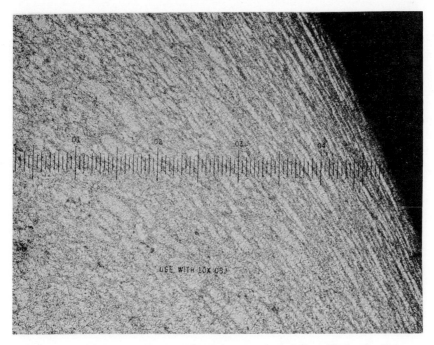

Figure 7. As-forged condition of P/M pinion gear on pitch line. Nital etch. 100x.

371

The last car test was the rock-cycle test—a jarring, full-throttle forward and reverse procedure, causing high-impact loading on differential gears. This is an extremely severe test on drive components and transmissions. One hundred rock cycles per gear set is a complete test, and twelve rock-cycle tests have been run on forged gears. All tests were positive [1].

The feasibility program on the pinion made from AISI 4023 is presently coming to a close. There were no different problems in forging this alloy as compared to the 4620. These gears have responded to heat treatment excellently, and have passed all physical and metallurgical tests. They are presently being introduced into a car-test program for final acceptance.

The AISI 8627 pinion program is presently being carried out. This alloy has not responded to heat treatment as well as the gear machined from wrought stock. The oxide content of this powder is somewhat higher than the 4620 and 4023, and the flowability is less. These problems are presently being investigated.

Side Gear

There are two P/M forging programs on side gears being conducted presently at Cincinnati Incorporated. The alloys for these two side gears are AISI 1522 and 4620. The gear will be forged with internal splines in the bore and to the shape shown in Figures 3 and 8. The only finish machining necessary will be the removal of two or three thousandths of an inch from the back face of the flange and the O.D. of the hub. Tolerance control has indicated that possibly one or both of these operations may be eliminated.

The designing of the side-gear preform became more complex than that of the pinion. Several preform shapes were tried before arriving at the final correct shape, as shown in Figure 8. The major problem was that of controlling metal flow in order to avoid forming laps in the forging gear. The gears are presently undergoing metallurgical tests, and the results look very favorable.

Other Applications

There are other applications which Cincinnati Incorporated and its customers are investigating, and there are many other applications which have been reported in current literature.

Buick Motor Division of General Motors Corporation has publicly announced that they are ready to introduce a 3.3-pound P/M forged input ring gear for their automatic transmission. The P/M forging saves 1.7 pounds of material per part over the conventionally forged part.

Figure 8. Side gear preform and P/M forged side gear.

Figure 9. P/M forged cam stator clutch.

Other transmission parts which are currently being developed are the cam stator clutch (Figure 9) and the sprague clutch. The alloy currently under investigation for these applications is 4620. These parts will have to withstand extremely high bearing loads on the cam surface of the I.D.

Other high-performance automotive applications include connecting rods for the engine, aluminum connecting rods for the air-conditioning compressor, bearing races, and several universal joint parts.

Summary

The P/M approach to forging has been shown to have its own process variables which most certainly set it apart from conventional forging. However, the ease of control of these process variables and the ability to form precision shapes at high production rates has been demonstrated for several high-performance applications.

References

1. Lusa, G., "Differential Gear by P/M Hot Forging," International Powder Metallurgy Conference, New York, July 1970, in *Modern Developments in Powder Metallurgy*, Vol. 4, *Processes*, Henry H. Hausner, ed., New York: Plenum Press (1971), 425.
2. Halter, R.F., "Pilot Production System for Hot Forging P/M Preforms," International Powder Metallurgy Conference, New York, July 1970, in *Modern Developments in Powder Metallurgy*, Vol. 4, *Processes*, Henry H. Hausner, ed., New York: Plenum Press (1971), 385.
3. Bargainier, R.B. and Hirschhorn, J.S., "Forging Studies of a Ni–Mo P/M Steel," Metals Engineering Congress, AMPI–ASM Powder Metallurgy Conference, Cleveland, Ohio, October 1970, in *Fall Powder Metallurgy Conference Proceedings*, New York: MPIF (1970), 191.
4. Cook, J.P., "Degradation of P/M Forging Preforms During Interim Exposures in Air," *Fall Powder Metallurgy Conference Proceedings*, New York: MPIF (1970), 237.
5. Pietrocini, T.W., "Hardenability of Hot Densified Powder Metal Alloys," 9th Conference of Metallurgists, sponsored by the Canadian Institute of Mining and Metallurgy and the Canadian Section, AMPI, Hamilton, Ontario, 25 August 1970.

Discussion

A High-Temperature Engineering Ceramic (Si_3N_4)
By Powder-Metallurgical Methods

MICHAEL W. LINDLEY   R. NATHAN KATZ
Admiralty Materials Laboratory   *Army Materials and Mechanics Research Ctr.*
Poole, Dorset, England   *Watertown, Massachusetts*

Silicon nitride (Si_3N_4) has been seriously investigated and used as an engineering material in the last few years. The material is produced in two principal forms; hot pressed [1] and reaction bonded [2,3]. Reaction-bonded silicon nitride in essence is not a single material but encompasses a spectrum of materials (like steel) with various property characteristics, and with an ability to be fabricated by differing processing techniques. The wide range of properties available to the designer/engineer results from these varied fabrication methods and subsequent processing conditions. Both forms of Si_3N_4 exhibit exceptionally good thermal shock resistance for a ceramic material, relatively good oxidation resistance, and good erosion/corrosion resistance [4]. These properties have led to consideration of silicon nitride ceramics for applications such as gas turbine combustors, rotors and stators [5], missile heatshields and nozzles [6], gas bearings, and furnace components.

While the highest strength silicon nitride is produced by hot pressing techniques, this brief review will limit itself to the discussion of reaction-bonded silicon nitride. It is most appropriate to discuss reaction-bonded Si_3N_4 ($RBSi_3N_4$) in a book on powder metallurgy, since this unique material represents a true interface where the technologies of powder metallurgy and ceramics merge. Reaction-bonded silicon nitride is made by producing a preform of silicon-metal powder by any one of several techniques (isostatic compaction, flame spraying, slip casting, or injection molding with a polymeric carrier). A replica of a complex shaped metal gas-turbine combustor, fabricated in silicon by the flame-spray technique is shown in Figure 1. The porous silicon-metal compact is then reacted in an atmosphere of nitrogen, and $RBSi_3N_4$ results. Thus, the ceramist literally works with powder metallurgy up to the point of sintering and then, instead of sintering in a neutral atmosphere, reacts the metal powder with the atmosphere during the firing stage to produce a ceramic. The several truly unique aspects of both the reaction-bonding process and the material which results will be discussed below.

The unique aspect of the reaction-bonding process in silicon nitride is that the dimensions of the original silicon-metal preform are preserved to within 0.1 percent. This means in practice that complex shapes can be machined from a

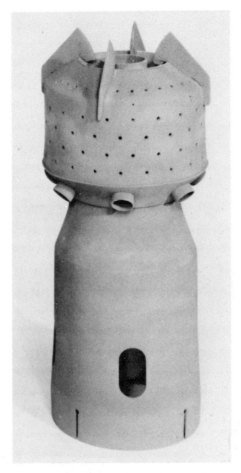

Figure 1. Flame-sprayed silicon metal gas-turbine combustor, prior to nitridation. (Courtesy of Admiralty Materials Laboratory.)

green silicon compact or, more preferably, from a lightly reacted compact (10-percent nitrogen uptake) known as "bisque fired" material. Such machinable compacts can be worked with conventional steel tools. Most flame-sprayed or molded silicon shapes require essentially no machining (although the as-sprayed silicon piece is usually robust enough to be machined without the need for an initial nitrogen firing). Since the fully nitrided piece preserves the dimensions of its precursor, costly machining of hard ceramic components is eliminated.

When silicon reacts with nitrogen to produce $RBSi_3N_4$, a 23-percent volume increase results. To accommodate this volume increase with only 0.1-percent

gross dimensional change requires that a substantial portion of the Si_3N_4 formed be formed in the void space. In fact, as will be shown below, the first Si_3N_4 to form occurs by whisker, platelet, or other intravoid growth of Si_3N_4. Since the attainment of a homogeneous piece of Si_3N_4 requires that nitrogen gas have access to all volume elements of the body, a continuously connected pore network is required. To insure connectivity of a second phase in a microstructure, 15 ± 3 volume percent of second phase (in this case pores) [7,8] is required. In practice, a silicon compact of about 60-percent green density yields an $RBSi_3N_4$ compact of about 75-77 percent density. Although high green densities and resulting high $RBSi_3N_4$ densities can be achieved in the flame-spray deposition

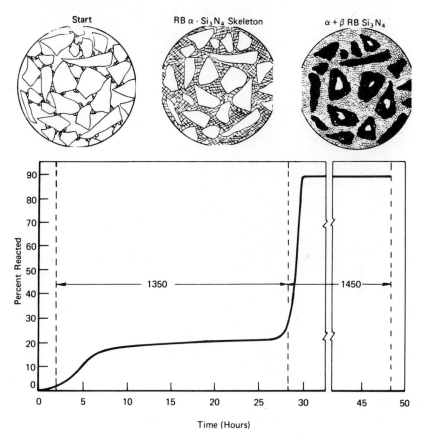

PERCENT REACTION VERSUS TIME FOR SILICON COMPACT REACTION
SINTERED IN FLOWING NITROGEN

Figure 2. Schematic of the reaction-bonding process as a function of time and temperature. The illustrations at the top represent the microstructure development at approximate positions along the nitridation curve.

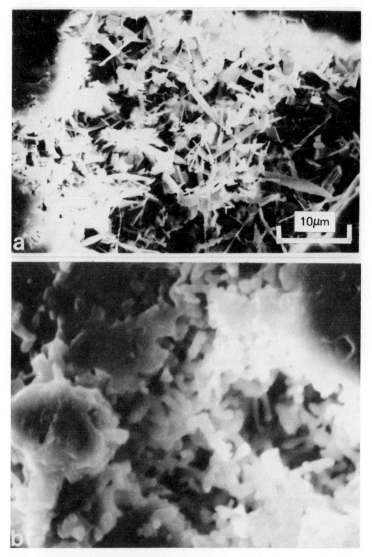

Figure 3. Scanning electron micrographs of duplex microstructures in RBSi$_3$N$_4$. (a) Cold pressed and reaction-bonded material [11]. 2000x. (b) Injection-molded and reaction-bonded material. 4000x. Note differences in whisker phase morphology in 3(a) and 3(b).

process, the highest green densities (excess of 1.9×10^3 kg/m^{-3} silicon green density) cannot be fully utilized since insufficient pore connectivity precludes totally reacting the silicon matrix. Silicon nitride occurs in two crystallographic forms; α-Si$_3$N$_4$ (actually an oxynitride Si$_{11.5}$N$_{15}$O$_{0.5}$), and β-Si$_3$N$_4$, both of which are hexagonal [9,10]. The α-form apparently occurs first at low temperatures (high O_2 potentials) [9] and starts filling the void space. This occurs at nitriding temperatures ranging from 1100–1400°C, and is illustrated schematically on the typical nitriding sequence shown in Figure 2. At the next stage of the nitriding sequence, the silicon metal is taken above its melting point and reacts with nitrogen to form both β- and α-Si_3N_4. The compact retains its structural integrity in the presence of liquid silicon metal because of the skeleton of α-Si_3N_4 which has been formed. An alternate hypothesis of buckling and exfoliation of α-Si_3N_4 films below the melting point of silicon can also account for this duplex structure, which is shown in Figure 3 [10].

The kinetics of the reaction-bonding process are being investigated by various workers but as yet are not well defined. The investigations which have been carried out demonstrate that particle size and purity are extremely important [10,11]. Iron impurities at the 1-percent level greatly accelerate the reaction to $RBSi_3N_4$. Initial powder particle size has been shown to effect final pore size [12] and thus mechanical properties of the final material, and also influences reaction rates while forming $RBSi_3N_4$ [11].

The mechanical properties of reaction-bonded silicon nitride are also rather interesting. Material with approximately 20-percent porosity has been obtained, with reported bend strengths of 43,000 psi (about $300MN/m^{-2}$). Since fully dense hot pressed Si_3N_4 has a bend strength of about 115,000 psi (about $800MN/m^{-2}$), the relatively high bend strength of the reaction-bonded material with 20-percent porosity is unusual. One may speculate that the answer may lie in duplex microstructure. There may be some reinforcement of the material by the α-Si_3N_4 whiskers formed *in situ*. If this should prove to be the case, a new class of composite materials may be considered—"autocomposites." However, much work needs to be done to elucidate the structure–property relationships in this material.

Acknowledgment

This article is published by joint permission of the Army Materials and Mechanics Research Center and Ministry of Defense U.K. (Navy Department).

References

1. Lumby, R.J. and Coe, R.F., "The Influence of Some Process Variables on the Mechanical Properties of Hot Pressed Silicon Nitride," *Proc. Brit. Ceram. Soc.*, 15 (1970), 91.
2. Brown, R.L., Godfrey, D.J., Lindley, M.W. and May, E.R.W., "Advances in the Tech-

nology of Silicon Nitride Ceramics," Special Ceramics Symposium, 14–16 July 1970. Stoke-on-Trent, England.

3. Parr, N.L. and May, E.R.W., "The Technology and Engineering Applications of Reaction Bonded Silicon Nitride," *Proc. Brit. Ceram. Soc.*, 7 (1967), 81.

4. Godfrey, D.J., "The Use of Ceramics in High Temperature Engineering," *Metals Mater.*, 2 (1968), 305.

5. McLean, A.F., "The Application of Ceramics to Small Gas Turbines," ASME Publication 70–GT–105, May 1970.

6. Godfrey, D.J., "The Fabrication and Properties of Silicon Nitride Ceramics and Their Relevance to Aerospace Applications," *J. Brit. Interplanet. Soc.*, 22 (1969), 353.

7. Cahn, J.W., "Phase Separation by Spinodal Decomposition in Isotropic Systems," *J. Chem. Phys.*, 42 (1965), 93.

8. Bishop, G.H., Quinn, G.D. and Katz, R.N., "Connectivity in Graded Two-Phase Structure," *Scripta Met.*, 5 (1971), 623.

9. Grievson, P., Jack, K.H. and Wilde, S., "The Crystal Chemistry of Ceramic Phases in the Silicon–Nitrogen–Oxygen and Related Systems," Symposia of BCRA on Special Ceramics, Stoke-on-Trent, England, 11–13 July 1967, in *Special Ceramics*, Vol. 4, New York: Academic Press (1968), 237.

10. Thompson, D.S. and Pratt, P.L., "The Structure of Silicon Nitride," *Science of Ceramics: Proceedings*, Vol. 3, G.H. Stewart, ed., New York: Academic Press (1967), 33.

11. Messier, D.R. and Wong, P., "Duplex Ceramic Structures–Interim Report No. 1: Kinetics of Fabrication of Silicon Nitride by Reaction Sintering" Army Materials and Mechanics Research Center, Watertown, Massachusetts, Technical Report, AMMRC TR–72–10, March 1972.

12. Evans, A.G. and Davidge, R.W., "The Strength and Oxidation of Reaction-Sintered Nitride," *J. Mat. Sci.*, 5 (1970), 314.

20. The Viability of an American Powder-Metallurgy Industry

K. H. ROLL
Metal Powder Industries Federation
American Powder Metallurgy Institute

ABSTRACT

After World War II enthusiasm ran high, if not rampant, for powder metallurgy's prospects. Since then, its evolution through many triumphs and tribulations has sobered the attitudes of many of those in, as well as outside, the P/M industry. So what are its prospects, its problems now? Is it a viable industry in light of today's economic and ecological trends? What influences, including those outside of the United States, will affect the future of the American P/M industry? In covering these and other vital points, the author offers a definitive, full-perspective analysis of America's P/M industry and its future.

Introduction

Is P/M any different than any other technology? Is it any different than any other business venture in metalworking? In some respects the answer to both questions must be "no." In other respects it is a definite "yes." P/M *is* different. And because of this difference it is worthy of a closer examination as to its prospects in a competitive world, a world shrinking so rapidly in the communication of its ideas while expanding so rapidly in the scope of its technological achievements that one cannot consider powder metallurgy as a separate entity—or even as the exclusive domain of any one nation. Yet to examine it in context, one must limit principal observations and conclusions to the American P/M industry in its own competitive society, the viability of an American P/M industry in an international industrial society, and the viability of an American P/M industry in a capitalistic system of free enterprise. This is what we will attempt to do.

Three Broad Directions of Growth

For many years I have described the progress of P/M as following two fairly distinct directions: one, where it offered another method of making a metal shape that could be made by any of several conventional metalworking techniques; where we were in head-to-head competition and our ability to offer cost savings coupled with good performance was our primary advantage. I will discuss this direction and its viability and prospects in more detail in a moment.

The other direction of P/M progress has been where the process itself offers unique advantages and results in products which cannot be made, or are impractical to make, by any other method of manufacture. Sintered metallic friction materials and cemented carbides are examples. Of course, these products are competitive within their own industry as well as with other ways of performing the function, but they are unique structures of materials engineering. There is no other way to make a cemented carbide, for example, except by starting with metal powder and applying P/M techniques. A P/M gear or cam, on the other hand, could easily be made by machining from solid bar stock.

I prefaced the above by stating that this was my view for many years—that P/M was developing along these two more or less distinct lines. Today, however, I believe we can now see the evolvement of a third line of progress: the merging of P/M techniques with conventional metalworking systems, where metal powders are the starting materials for common metalworking. Often, the benefits are high performance or improved economies, or both. The marriage of forging and P/M is a prime example. Extrusion of P/M billets is another. Roll-compacting of metal powder could also be considered in this category.

This third direction of P/M evolution has the potential to become the most important of the three in terms of volume—dollar as well as tonnage. At this point in time its parameters are still undefined, but in another sense it is easier to see where it could go than the previous direction. Unique applications tend to become a function of need, and prognosticating tomorrow's technological needs can be dangerous indeed—except for certain basics. They are needs that will undoubtedly become more and more difficult to meet through today's practices, because they will involve higher temperatures, lower temperatures, higher stresses, and greater concern for both labor and materials savings.

Nevertheless, let us examine the prospects for this direction of powder metallurgy, at least in general terms. When we entered the space age we also entered a new era of materials engineering. It could be said that we conquered space because we were able to change our approach to materials utilization. We created the material to meet the need, rather than limiting the application to the materials available. I relate this to P/M because the powder metallurgist has been, and still is, in the forefront of the materials revolution. He has made it possible for the materials and design engineer to broaden his horizons far beyond the handbooks by enabling him to combine properties of materials that would otherwise be unheard of. For example, alloys that might normally be immiscible, combinations of metals and nonmetals possessing the characteristics of each, whiskers or fibers oriented in a metal matrix, and so on.

Perhaps the simplest way to illustrate this contention is to compare the powder metallurgist to the organic chemist. I am sure everyone today realizes what has been accomplished by the organic chemist and how much he has contributed to our progress because of his ability to disassemble the molecule and then reconstruct it to create infinite combinations with properties tailored to the application. Our entire petrochemical and plastics industries are based on this facility.

Is not this, in a sense, what the powder metallurgist is doing by breaking his metallic materials down into discrete elemental particles and then recombining them under precisely controlled conditions? In this way he can form materials systems of combinations and characteristics otherwise unattainable. Cemented carbides and sintered friction materials are such systems.

Perhaps, as in the case of the early organic chemist, it could be said that the potentials for this concept have not even begun to be realized. There is no end in sight for the ingenuity of man's ability to meet his needs. Nor is there any apparent limit to man's needs. No matter what we do or where we go—outer space or inner space—we will need materials to support us, convey us, protect us. Powder metallurgy will surely play an increasing role in meeting these needs.

So far, with minor exceptions as illustrated above, we have limited our thinking to the tried and true, to the alloys and combinations of materials that have emerged through the traditional routes. In metals, we have been bounded by the ordinates of the phase diagram and limitations of the liquidus.

The first signs of breaking through this barrier of convention are beginning to appear. P/M forging research is moving toward alloy systems that are impractical for conventional forging. The application of powder metallurgy to tool steels, and the creation of a whole new family of P/M tool steels, is another example of the embryonic awareness of the potential for powder metallurgy.

Thus, it appears to me that P/M is beginning to enter into a new era, technologically speaking, where we expand the unique areas of applications and merge with traditional metalworking.

Evolution of the P/M Industry in the United States

Let us leave the future for the moment, and take an introspective look at those influences that have exerted themselves on powder metallurgy's evolution. Why has it reached its present state of development? Evolution, by definition, implies a gradual growth; one step forward after another, each imperceptible in its effect. While this undoubtedly is true in the evolvement of today's metal-powder producing and consuming industries, I believe we have been strongly affected by certain specific events which perhaps could be considered "breakthroughs." Four of these are worthy of examination. It is important to examine them because of their possible bearing on the future as well as their fragility: after all, we could slip back unknowingly.

First Breakthrough

First and foremost is the unveiling of P/M as a modern technology, subject to the same pluses and minuses that every emerging technology faces in the process of winning acceptance and establishing credibility. It is *not* black magic. The powder metallurgist does not mutter incantations over a pile of secret powders to convert them mysteriously into products capable of out-performing wrought metal—and all at a small fraction of the cost. "Tell it like it is," is the popular phrase. It took many years to agree to do this, and the result of the decision to do so was an eruption into a traumatic confrontation that produced a serious schism within the industry itself—but it was done. "Give them the facts and tell them enough about P/M so they can compare it on equal terms with products of other processes and then make an intelligent decision to use it where it belongs and leave it alone where it does not belong." That is the philosophy under which the industry operates today. It is now accepted for the most part, but it was a breakthrough eight years ago.

Second Breakthrough

The second major breakthrough occurred when the P/M parts industry decided to face up to the need for realistic standards. Any materials standard is at least a compromise and at best a realistic performance indicator to which both manufacturer and consumer can subscribe with confidence. For too many years P/M materials standards were compromises that lagged far behind the industry's capabilities. In the early days, with low performance adopted as standard, it simplified the selling of higher performance parts. But soon designers who worked from these standards and who were unaware of the industry's real abilities shied away from P/M. Many applications that easily fell within the performance limits of conventional powder metallurgy were not being considered because of outdated and downgraded standards.

It was only three years ago that the Standards Committee of the Powder Metallurgy Parts Association agreed to update and upgrade its standards. The new data were generated and released in the form of revised MPIF P/M Materials Standards and Specifications No. 35. Additional new data will soon be added, and it is hoped that this will mark the beginning of a continuing system of review and improvement so that the designer may confidently incorporate P/M products in his plans. To remain viable in the face of constantly improving competing materials demands continuing progress in P/M materials and the standards that represent them, coupled with an awareness of quality and performance reliability and a willingness to enforce it.

Third Breakthrough

The third breakthrough came when the custom P/M parts industry decided that the in-plant manufacturer was not such a bad bedfellow after all. He could not only contribute to the industry technologically but he could also help open up more markets instead of swallowing up existing ones.

"Make or buy" is a time-honored choice of the American businessman, but for many years fierce resentment typified the attitude of the P/M parts maker who discovered one of his customers deciding to "make" instead of "buy." To examine the reasons for this situation would serve no purpose here. Suffice it to say that custom and in-plant P/M parts makers both serve a need in the evolution of a viable P/M industry. They tend to keep each other on their toes—technically and economically. If the parts bought are not good enough, or are too costly, the in-plant manufacturer has the alternative of making his own. If the parts involve too high a risk for a custom maker to justify or afford, the in-plant maker usually has the resources—money, manpower, and equipment—to do the job. Often, a new application results and opens up further opportunity for P/M from which everyone benefits.

By the same token, the custom parts maker contributes from a vast storehouse of diversified experience and a variety of process equipment. Certainly the in-plant maker is wise to buy whenever feasible and take advantage of the skills that the established custom P/M parts maker has to offer.

What really counts is that today, the in-plant and custom P/M parts makers are linked together through MPIF without violating their independence, while enabling them to work cooperatively toward the creation of better materials, better processing techniques, better standards, and a better P/M industry.

I have made several references to the Metal Powder Industries Federation. Modesty prevents me from labeling it a breakthrough, but few would deny that the establishment of MPIF was a unique concept in the organization of a total industry complex. The entire spectrum of powder metallurgy is represented in the ranks of the Federation—all the metal powder producing and consuming industries and all the ancillary support industries, including process equipment makers, plus a direct coupling with the industry's professional society, the American Powder Metallurgy Institute. MPIF has been, and will continue to be, the strongest force in the development of a viable P/M industry, unlimited with respect to national boundaries.

P/M and Forgings—The Fourth Breakthrough?

A discussion of breakthroughs could not overlook P/M forgings, though only history will show whether this is a breakthrough or a missed opportunity. On other occasions I referred to the marriage between P/M and forging. The baby has now been born, and will undoubtedly grow up and turn out all right—but meanwhile its parents worry because much of the maturing process is out of their hands. And worry they should, because they cannot even find a good name for it.

In our desire to associate ourselves with forging and to invade the market domain held by this venerable process, we may have overlooked resistance not only from the forgers themselves but from their customers as well. And these are the people we must cultivate and convert. Their resistance stems from our customary problems—education and public relations. They are not yet convinced that we can make good on our statements. "Let's wait and see what happens in automotive before we take a chance," they say. Influencing their attitude is the fact that we are neither fish nor fowl at this point. We must somehow establish our own identity as a new and unique product with attributes of both P/M and forging but yet totally distinct from either.

Prospects for P/M Parts Manufacture

In our zeal for the new and exciting we would be terribly remiss to ignore the real strength of our industry: the manufacture of P/M parts and bearings. Therefore, let us examine more closely the influences affecting this facet of our business and its prospects for the future.

Page upon page of observations and commentaries about the markets for P/M parts has already been written. There is no need to say more. Instead, I would rather address myself to some comments about technological progress in the manufacture of P/M parts.

The progress of powder metallurgy was greatly enhanced when compacting-press manufacturers recognized that the mere substitution of a pill press for compacting metal powders was by no means the ultimate and began to design consolidation systems expressly for metal powders. There is no question in my mind that much of the progress our industry has made today in terms of market penetration, size of parts, properties, and configuration can be attributed in large measure to the research and development efforts and progressive attitudes of the compacting-press manufacturers. Nor can we ignore the forces of competition. After all, P/M represented a new market and one with an attractive potential; therefore, many companies—and many nations—entered the race. Competition was enhanced further by the attitude of the custom P/M parts maker who said, "If you won't make the equipment I need I'll build it myself." And in many cases did. Certainly the powder producer has also contributed to the advancement of consolidation techniques by improving his powders, making them more compressible and more consistent in quality, and at a price that enabled the parts maker to compete effectively with other metalworking industries.

But the consolidation step of the powder-metallurgy process, if one ignores the tooling, is relatively uncomplicated. It occurs at room temperature. If mistakes are made, the losses are not significant and adjustments can be made, or, at worst, the green parts can be converted back into powder.

Unfortunately, such is not the case in sintering. Mistakes here can be costly. Sintered scrap is almost useless. There are more variables to deal with: time, temperature, atmosphere, lubricant burn-off, cooling rates, and heat-up rates. The equipment itself is costly and the controls, largely because of special atmospheres, are intricate and complicated. In compacting, the only energy input is to operate the press, while in sintering, energy not only moves the parts through the furnace but is needed to supply heat for the process and a constant supply of expendable gas for maintaining a controlled atmosphere.

Need for Innovation in Sintering

Does it seem strange, then, that by far most of the innovation in the powder-metallurgy process has come from the consolidation aspect and extremely little from sintering?

Let us examine this thesis a little further. The primary market for the furnace manufacturer was heat-treating and brazing, and probably still is; therefore, the sintering market was not sufficiently attractive to justify much effort in converting a brazing or heat-treating furnace into a sintering furnace. There is another reason: the typical P/M parts maker often found it relatively easy to construct his own conversions, thus minimizing incentive to the custom furnace manufacturer. Also, the custom parts maker would frequently purchase a new compacting press with the idea of amortizing its cost through one or two new jobs—but rarely could one write off the cost of a sintering furnace on this basis. Besides, it was comparatively easy to increase the production rate on an existing sintering furnace by running it around the clock or perhaps even cutting corners on time and temperature.

Whatever the reasons, it is still apparent that with few exceptions the sintering furnace industry as a whole has not as yet made any basic contributions to the advancement of the powder-metallurgy process. The question that could be asked is: "Can they?" Perhaps the nature of the process is such that no further innovation is possible, except for minor engineering and materials handling improvements. However, I am not sure I would subscribe to that argument, for I believe there is fully as much opportunity for innovation in sintering as there was in consolidation. For example, it would seem obvious that we are not yet taking full advantage of the fact that a P/M part is at an elevated temperature and in a neatly controlled atmosphere during sintering and, as such, fully receptive to further consolidation or reforming. We have not yet taken full advantage of the ability to heat-treat parts during the sintering phase. The advent of P/M forging may play an important part in sintering. It is not inconceivable that this concept might virtually eliminate the traditional sintering step. For example, a green compact could be induction-heated and forged under oxidation-controlled conditions to produce a fully dense sintered product.

I am not suggesting here that the answer lies in the utopian concept of hot pressing, wherein we combine consolidation and sintering into one operation. This is an approach that works in some instances and is necessary in others, but is faulted by the fact that the powder metallurgist enjoys, and more often needs, the ability to separate his two basic processing steps—consolidation and sintering —because in each of these he faces variables which are far easier to control and manipulate separately, rather than in combination.

In short, there appears to be much room for advancement and opportunity for innovation in the sintering phase of powder metallurgy. How and when this will come about I dare not venture, other than that it is inevitable—necessity still being the mother of invention.

World-wide P/M Viability

Now I would like to step back and examine our industry from a broader perspective, for I do not believe we can consider the viability of an American P/M industry out of context with the viability of a world-wide P/M industry. We are all familiar with the European origins of modern powder metallurgy and with the progress that has been achieved here in the United States. Powder metallurgy has contributed substantially toward resolving the needs of a mass-production industrialized economy coupled with a high and rising standard of living. Additional factors in our favor have been the driving forces of free enterprice and the presence of individualists with pioneering spirit and the courage to carve out a new technological frontier.

But let us consider other industrialized nations around the world, because I believe that in some respects they are going to outstrip us in taking advantage of what powder metallurgy has to offer. First of all, they have the benefit of being able to foresee what can happen as their economy and standard of living advances. They do this by merely observing what is going on over here. Thus, they can avoid our mistakes and exploit our advances. Second, they often have the advantage of a beneficent government. The governments of Japan and Canada, for example, nurse and encourage their powder-metallurgy industries. Third, some of them have reaped the strange advantage of serving as the battlefields of World War II. We lost men and material, but not one of our plants was destroyed. In other words, we are hampered by many, many obsolete manufacturing facilities crippled by time, technical obsolescence, labor-union activity, and unrealistic depreciation allowances. Japan, Europe, and Russia are not.

We cannot afford dramatic innovation if it means making obsolete an existing facility. The roll-compacting of steel powder to make high-quality steel strip represents dramatic innovation, but we will not see it as a commercial reality as long as our existing strip mills are still around or as long as the economics of obsolescence cannot somehow be circumvented.

So what will happen in those overseas countries which have recognized the viability of our American P/M industry? In some instances they will merely continue to follow in our footsteps. But others—Japan and Russia, for example—are in an excellent position to make a quantum leap: to bypass or hurdle the accepted traditional methods and go directly to P/M—with obvious savings in time, investment, and people. Why build huge blast furnaces and open hearths and slab-rolling mills when with an iron-powder producing facility one can go directly into steel strip by roll-compacting; into vacuum-melt equivalent quality steels by electro-slag remelt with powder; into alloy P/M slabs and ingots free from segregation, porosity, and nonhomogeneity; into tool steels and superalloys by isostatic forming, forging, or extrusion of P/M compacts?

The only thing saving us is that we have the markets here, the demand for better superalloys and tool steels. And we have the pressure of increasing labor costs, hence the need to minimize labor input.

Paralleling this advantage of P/M for combatting labor inflation is an increase in the availability of laborers. "The labor force is swelling at a rate double that of the past decade," says *Iron Age*, "and in order to maintain full employment in the American economy we must exceed the highest national economic growth rate that we have ever experienced in the past." How do we do it? *Iron Age* editors believe it is foreseeable that through the unprecedented need for growth, meaningful depreciation reform will finally become a reality in this country. The compelling urgency for creating jobs at double the past rate means more tools of production will be needed. New tools, efficient tools—those tools that spell high productivity. And this is where P/M shines. We are ideally suited to mass-produce the components that go into the durable goods whose manufacture offers America the opportunity for full employment. Our citizenry will continue to demand more time for the pursuit of leisure and more of the good things of life—most of which utilize or should utilize P/M products.

Ecologically Clean

Another name in the game is "ecology." It is our good fortune that powder metallurgy is being heralded as an ecologically acceptable manufacturing method. There has been so much talk and attention directed to ecology these days that perhaps we should not pass over the subject too lightly. It can play a major role in the evolution of P/M, even if the current fanaticism about ecological rape diminishes to mere hysteria. P/M is not only an "ecologically clean" process in that it does not create pollutants—air or water—but it also does not generate scrap or waste materials. As a matter of fact, many of the present methods of manufacturing metal powder are recycling techniques that utilize scrap as raw material. Most of today's atomized iron powder is made from scrap iron. Hydrogen-reduced iron powder is made from mill scale, a waste product of steel rolling mills. Copper powder is made from scrap copper or from wire-drawing scale, or from mine waters that would otherwise pollute streams. Metals in powder form can be recovered from sources, scrap or natural, that could not serve economically as raw materials for conventional metal-recovery systems. Some day, chemical mining of low-grade copper will no doubt be commercially feasible because of copper-powder metallurgy.

In the sintering operation we utilize "clean" heat—either gas or electricity—and generate water vapor and carbon dioxide, both vital to biological systems. The tiny fraction of metallic stearate die lubricants volatilized in the sintering operation are either trapped in the furnace or, if vented, are colorless, smokeless, and harmless. The atmosphere in, around, and above a P/M parts plant is far cleaner than that inside a modern jet aircraft filled with passengers fouling the air with cold sweat, martini vapors, carbon dioxide, and tobacco smoke.

The P/M industry is also benefiting from an "ecological windfall"—the high costs of converting older foundries in order to comply with new safety, health, and antipollution regulations—thus tipping the economic balance in favor of P/M in such areas as cast versus P/M connecting rods.

Profitability is the Key to Viability

But nothing that has ever been said about the glowing prospects for this industry will come to pass unless it is able to achieve adequate profitability. Without sufficient return on investment there can be no innovation, no new-product development, no effective promotion, no research to meet tomorrow's needs—but above all, no incentive to remain in this business, much less invest in it.

I am fearful that this industry may never recover from the effects of the Great Recession of 1970 unless collectively and totally it can mend its business fences and return at least to the point where it left off before the recession, and before excessive competition within its own ranks destroyed its ability to operate at a profit. The primordial urge for mere survival by taking on any job at any price is not easy to overcome or recuperate from. A return to normalcy must be made. Not to do so spells disaster in every sense of the word. The P/M parts industry runs the grave risk of being left behind; of becoming metal converters and nothing more; of being unable to innovate, of being beaten to death by each other. What an ignominious end for such a promising industry!

Who are the perpetrators of this genocidal self-destruction? No, they are not the shadowy competitors who skulk in the dark and steal one's business when one is not looking. They are not the little fly-by-nighter. P/M is no longer a fly-by-night business. The back-alley shop? Perhaps. But do not forget that some of today's most successful companies started out in just this manner. They had guts and skill and salesmanship. Today, it takes at least one more condition: money. Lots of it. Plus the ability to manage it.

No, I am afraid the guilty ones are some of the biggest, some of the smallest, some of the newest, some of the oldest. But they all have one trait in common— they are convinced of their own maiden purity; they are convinced that their competitor is a stupid business barbarian who does not know metal powder from face powder and who hires as salesmen grade-school dropouts who now hold advanced degrees in larceny and persiflage.

So what is the answer? Purge? Good idea. But who? Quite possibly, everyone could be laid out for passage through that great sintering furnace in the sky! Besides, who is to judge the guilty ones? Everyone is guilty of business stupidity at some time or other.

How about education as a cure? It is slow, but it does work. The only problem is how to get the miscreants to come to school or join the church. How about getting prices up and keeping them up by mutual agreement; a "gentlemen's agreement" sworn to over multiple martinis but forgotten as soon as the hangover—or as soon as violation of the pact is suspected? It won't work. In a custom product industry such as this, any collusionary attempts are doomed to failure from the outset because it does not market any products in common. It sells only skills and ability to use the manufacturing equipment it possesses.

It should be apparent by now that there is no magic answer to profitability in any business. It has to be a function of each company's manufacturing capabilities and each manager's business abilities coupled with the desire to make a profit. And herein lies the key: each man in the P/M industry must judge for himself whether or not he can make a profit. He must not worry about whether or not his competitor does. That is the competitor's concern. If everyone would decide what he considers to be a fair return on his investment and then go about getting it, the P/M industry will become profitable.

Action in Concert

Once the path to profitability is actively being pursued, then action in concert to exploit the market potential to its fullest must be taken by the three primary elements of the P/M industry—powder producers, custom P/M parts makers, and P/M equipment suppliers. I do not deny that this can be accomplished gradually on an individual company basis. In fact, each company in the P/M business must continue to exploit its own capabilities, develop its own specialties, establish clearly its raison d'etre. Action in concert—a trade-association activity if you will—cannot substitute for individual company initiative in the marketplace. It does only those things which can best be done through cooperative effort or which cannot be done alone. Classically, these are to gather statistics, to write standards, educate customers, and promote the industry and its products.

It is in the latter category that I believe the greatest good can be accomplished through effective, coordinated, cooperative action by a trade association. That it can be done and that it can develop markets and create a business potential has already been demonstrated by the Metal Powder Industries Federation. But the ultimate achievement—a sound business basis built on a universal belief in profitability and conducted in an atmosphere of cooperation, not collusion—has not yet been achieved in the P/M industry. And it will not be achieved in my considered judgment until all three elements cited above are willing to share equally in the creation and support of long-range programs designed to put P/M in every application where it properly belongs.

I have participated in the beginning of this triumvirate troika through the founding of MPIF. But this can only be the beginning. When it is accomplished, it may well be the most significant of all the breakthroughs in the progress of P/M and a milestone in the establishment of a viable American P/M industry.

Before leaving this subject, let me reiterate one point in my premise that perhaps might have escaped unnoticed or whose import might not have been apparent: "To share equally in the creation and support of long-range programs." By this I mean that the powder producers, the custom parts makers, and the equipment makers sooner or later must sit down together, decide what they want to do, how much money they need to do it with, get the money up by dividing it equitably, perhaps equally, among all participants, and then go ahead and do it.

These three are the direct beneficiaries of an expanded, healthy P/M market; these three therefore should bear the burden of bringing into being that which will benefit them most. No one can deny that a strong P/M parts industry automatically means a strong and innovative metal-powder and P/M equipment industry.

Summary

In summary, then, the *static growth* at best—or the decline at worst—of the American P/M industry will be dictated by (1) the attitudes and confidence of those now in the industry; (2) the amount of research and development effort expended; (3) the competitiveness of other existing techniques; and (4) the emergence of new techniques that might make powder metallurgy obsolete.

The *dynamic growth* of the American P/M industry will come about only if (1) adequate profitability can be achieved for all elements that comprise the industry—and that depends on a reasonable degree of price stability, *i.e.*, high enough for profit through volume but low enough to permit mass-market expansion in areas that now enjoy low costs; and (2) the willingness of the three basic elements in the industry—P/M parts makers, powder producers, and equipment suppliers—to share in the creation and support of long-range programs for developing and exploiting the markets for P/M products.

Everyone must be convinced that the P/M industry is indeed viable and that its potentials are greater today than its pioneering founders ever imagined.

Index